Ocean Engineering & Oceanography

Volume 28

The Ocean Engineering & Oceanography (OEO) series presents cutting-edge research and practical applications within the interconnected realms of ocean engineering and oceanography. This collection contains monographs, edited volumes, conference Proceedings, and lecture notes that delve into diverse theoretical and applied aspects of research in both ocean engineering and oceanography. Covered topics span various areas, such as marine electronic engineering, focusing on the development of electronic devices tailored for marine use; offshore engineering (or offshore construction), which revolves around the design of fixed and floating marine structures, pipelines, and other submerged equipment; and oceanography, addressing the study of physical, chemical, geological, and biological processes within the marine environment.

All books published in the series are submitted for consideration in Web of Science.

Han Zhang · Jie Zhang · Ruinan Lin · Yifei Wang

Submarine Pipeline Integrity: Assessment of Failure Modes and Advanced Evaluation Techniques

Springer

Han Zhang
CNOOC International Company, Limited
Beijing, China

Ruinan Lin
PetroChina Southwest Oil and Gasfield Company
Chengdu City, Sichuan, China

Jie Zhang
Southwest Petroleum University
Chengdu, Sichuan, China

Yifei Wang
Department of Mechanical Engineering
University of Alberta
Edmonton, AB, Canada

ISSN 2194-6396 ISSN 2194-640X (electronic)
Ocean Engineering & Oceanography
ISBN 978-3-031-92091-2 ISBN 978-3-031-92092-9 (eBook)
https://doi.org/10.1007/978-3-031-92092-9

This Springer imprint is published by the registered company Springer Nature Switzerland AG
The registered company address is: Gewerbestrasse 11, 6330 Cham, Switzerland

Preface

Submarine pipelines are critical global energy infrastructure, enabling the transportation of oil, gas, hydrogen, and other critical resources across extensive underwater distances. As offshore resource exploration continues to expand, the strategic importance of these pipelines grows accordingly. However, they operate in some of the most challenging environments on Earth, where they are subjected to a range of threats, including corrosion, mechanical damage, and seabed instability. These factors heighten the risk of pipeline failure, environmental harm, and substantial economic losses. Ensuring the safe and efficient operation of submarine pipelines demands a comprehensive understanding of both conventional steel and advanced composite materials, major failure modes, as well as the implementation of cutting-edge inspection, evaluation, and repair techniques.

Submarine Pipeline Integrity: Assessment of Failure Modes and Advanced Evaluation Techniques addresses these critical challenges directly. It represents the culmination of decades of research, field experience, and the latest technological advancements in submarine pipeline integrity management—from inspection protocols to innovative repair strategies. Tailored for professionals—including engineers, project managers, and researchers—this book offers advanced insights and practical tools essential for the design, maintenance, and repair of submarine pipelines operating under complex conditions. It ensures their reliability and durability in the face of increasingly challenging marine environments.

We, the authors, have drawn upon our diverse expertise in pipeline engineering, reliability engineering, and artificial intelligence to provide readers with the most up-to-date knowledge in the field. Our collective experience, shaped by Ph.D.-level research and extensive practical work, has enabled us to address both the foundational aspects of pipeline integrity and the latest advancements in inspection and repair technologies. Chapter 1 explores advanced inspection techniques, covering visual, non-destructive testing, and intelligent technologies. Chapter 2 focuses on assessing corroded pipelines using simulations and reliability analysis. Chapter 3 examines impact damage mechanisms, while Chap. 4 analyzes thermal buckling behavior. Chapters 5–7 cover composite pipelines, including their mechanical properties and failure mechanisms under complex loads. Chapter 8 addresses dent behaviors in steel pipelines, and Chap. 9 highlights innovative repair techniques and emerging trends.

This book stands out for its comprehensive coverage of both traditional steel and advanced composite pipelines. Unlike many publications that focus on a single material or provide limited inspection methods, it takes a holistic approach, offering advanced evaluation techniques for various materials and failure modes. From mechanical behavior analysis to cutting-edge AI techniques for predictive maintenance, the insights reflect the authors' collective expertise.

The first author, Han Zhang, extends his sincere gratitude to his research advisors, Dr. Zhigang Tian and Dr. Mingjian Zuo, for their unwavering support and guidance over the years. He is also deeply thankful to his family and close friends, Dr. Zheng and Ye Yuan, for their continuous encouragement and love. The second author, Jie Zhang, extends his heartfelt thanks to his family and the research team at Southwest Petroleum University for their support. The third author, Ruinan Lin, expresses his gratitude to his family and close friends for their love and encouragement. The final author, Yifei Wang, offers his sincere appreciation to his family and research colleagues for their unwavering support.

The authors wish to express their sincere gratitude to the editors for their exceptional collaboration and meticulous attention, which have been instrumental in bringing this book to publication.

Beijing, China — Han Zhang
Chengdu, China — Jie Zhang
Chengdu, China — Ruinan Lin
Edmonton, Canada — Yifei Wang

Acknowledgements This book has received financial support from the following projects: National Natural Science Foundation of China (52104223); Key Research and Development Program of Sichuan Province (2024YFFK0422). We would also like to thank the following individuals for their assistance:

Chapter 1
Hao Zheng, Faculty of science, University of Alberta, Edmonton, AB, Canada T6G 2G8

Chapter 2
Zhigang Tian, Department of Mechanical Engineering, University of Alberta, Edmonton, AB, Canada T6G 2G8

Mingjian Zuo, Department of Mechanical Engineering, University of Alberta, Edmonton, AB, Canada T6G 2G8

Chapter 3
Ke Song, School of Mechatronic Engineering, Southwest Petroleum University, Chengdu 610500, China

Zhiqiang Chen, School of Mechatronic Engineering, Southwest Petroleum University, Chengdu 610500, China

Chapter 4
Xin Lu, T School of Mechatronic Engineering, Southwest Petroleum University, Chengdu, China 610500

Chapter 5
Bofeng Liang, China University of Petroleum (East China), Qingdao, China 225319

Chapter 6
Te Hu, Key Laboratory of petroleum and natural gas equipment, Ministry of Education, Chengdu, China 610500

Chapter 7
Rui Yang, Oil and Gas Equipment Technology Sharing and Service Platform of Sichuan Province Chengdu, China 610500

Haoming Sun, School of New Energy and Materials, Southwest Petroleum University, Chengdu, China 610500

Chapter 8
Chuanjun Han, School of Mechatronic Engineering, Southwest Petroleum University, Chengdu 610500, China

Competing Interests The authors have no competing interests to declare that are relevant to the content of this manuscript.

Contents

About the Authors

Han Zhang, Ph.D. is a recognized authority in pipeline integrity management and reliability engineering, specializing in the structural reliability and failure assessment of both onshore and submarine pipelines. He earned his doctorate from the University of Alberta and brings extensive expertise across a broad range of areas, including pipeline mechanics, hydrogen energy transportation and utilization, advanced composite materials, machine learning, and maintenance optimization for energy and manufacturing systems. Dr. Zhang currently works at CNOOC International Company, Limited, where he applies his knowledge to real-world engineering challenges in offshore energy infrastructure. Dr. Zhang has made significant contributions to the energy sector, authoring two monographs and over 40 papers in prestigious international journals. His work seamlessly blends theoretical research with practical applications, driving innovation and advancements in both traditional and renewable energy industries.

Jie Zhang born in 1987, graduated from Shijiazhuang Railway University with a bachelor's degree in 2009 and obtained a Ph.D. in Mechanical Engineering from Southwest Petroleum University in 2016. He was a visiting scholar at the University of Alberta in Canada from December 2016 to June 2017. His research interests include the service safety of pipelines and pressure vessels, hydrogen energy equipment, and hydrogen safety. He has published four monographs and over 120 papers and has been granted 25 Chinese invention patents. He has been recognized as a reserve candidate for academic and technological leadership in Sichuan Province, a young talent of the China Association for Science and Technology's think tank program, and an outstanding young energy technology professional by the China Energy Research Association. He is listed among the World's Top 2% Scientists.

Dr. Ruinan Lin is a senior engineer at PetroChina Southwest Oil and Gas Field Company. He earned his doctorate from China University of Petroleum-Beijing in 2020 and has since focused his research on oil and gas storage and transportation engineering, safety engineering, and pipeline mechanics. Over the years, Dr. Lin has been actively involved in more than ten scientific research projects and has published 17 articles in reputable academic journals.

Yifei Wang is a postdoctoral fellow at the University of Alberta, with a research focus on health monitoring, reliability engineering, machine learning, and maintenance decision-making for energy and manufacturing systems. His research contributions have been recognized in several publications in peer-reviewed international journals.

Chapter 1
Advanced Inspection Techniques for Submarine Pipeline Integrity

1.1 Introduction

This chapter addresses the pivotal role of inspection in maintaining the integrity of pipelines laid in the challenging marine environment. Defects such as corrosion, cracks, and dents in steel pipelines, along with delamination, fiber breakage, and matrix cracking in composite pipelines, significantly compromise the structural integrity and operational efficiency of submarine pipelines, posing serious safety and reliability threats. Besides, defects in pipelines can also pose risks to the marine environment and human safety without proper maintenance strategies. This chapter looks into the significance of inspection methods and the latest developments in inspection technologies necessary for identifying defects and ensuring the ongoing, safe functioning of submarine pipelines. Figure 1.1 depicts a scenario in which a pipeline operator inspects a gas pipeline.

1.2 Inspection Methods

Over recent decades, inspection techniques used in underwater environments for both metallic and composite materials have significantly evolved. Such evolution reflects a growing demand and diversity for monitoring submarine pipelines [1]. Techniques such as ROV inspections, acoustic monitoring, and magnetic flux leakage exemplify the array of tools available for assessing the condition of submarine pipelines [2]. Selecting the most suitable inspection method is critical, as it significantly impacts maintenance efficacy. This decision requires a nuanced understanding of each method's strengths and limitations. Hence, a comprehensive evaluation of the pros and cons of common inspection techniques is vital for upholding the integrity and functionality of submarine pipelines.

H. Zhang et al., *Submarine Pipeline Integrity: Assessment of Failure Modes and Advanced Evaluation Techniques*, Ocean Engineering & Oceanography 28,
https://doi.org/10.1007/978-3-031-92092-9_1

Fig. 1.1 Gas pipeline inspection scenario

1.2.1 Visual Inspection Methods

Visual inspection techniques involve using optical tools and methods to observe the external condition of submarine pipelines directly. These techniques encompass a broad range of technologies, from basic visual checks performed by divers to sophisticated imaging systems mounted on remotely operated vehicles (ROVs) and autonomous underwater vehicles (AUVs) [3]. High-definition cameras, laser scanning, and sonar imaging are among the advanced tools that provide detailed visual data, allowing for the identification of structural issues such as corrosion, cracks, and biofouling. By providing clear, detailed views of the pipeline's exterior, visual inspection methods enable operators to detect and address defects early, minimizing risks and avoiding costly downtime [4]. The following outlines key visual inspection methods employed for submarine pipelines.

1.2.1.1 Remotely Operated Underwater Vehicles

ROV is a tethered underwater robot equipped with several high-resolution cameras and tools, such as chemical samplers and lighting systems. Operated remotely from the surface, it serves a pivotal role in executing a variety of underwater tasks, including pipeline inspections, repairs, and exploratory missions, at depths inaccessible to divers. Moreover, ROVs are capable of transmitting real-time data and visuals to the surface, enhancing the efficiency and safety of pipeline inspections. However, deploying ROVs for submarine pipeline management comes with challenges, including substantial initial costs for the system and the need for specialized

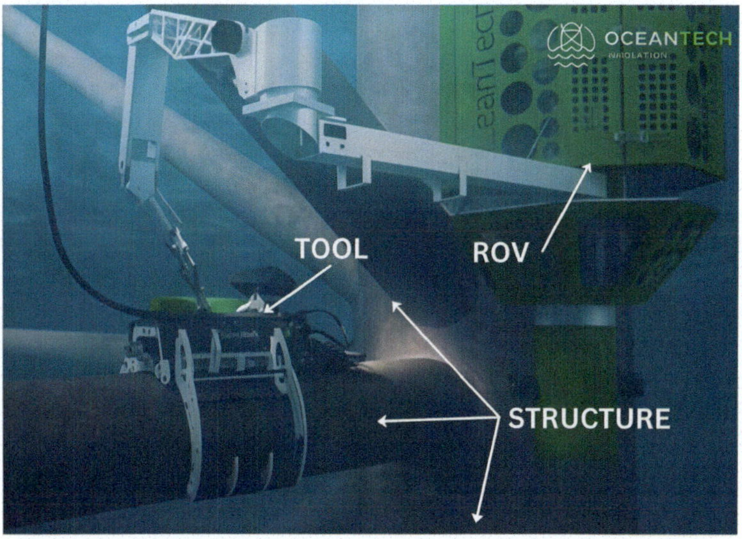

Fig. 1.2 An example of ROV [3]

operators. Environmental factors like strong currents, poor visibility, and weather disturbances, particularly tugging on the tether, can impact ROV effectiveness [5]. Figure 1.2 shows an example of an ROV used for submarine structures.

1.2.1.2 Autonomous Underwater Vehicles

AUVs are automatic robotic systems designed for underwater tasks such as mapping, surveying, and inspecting submarine pipelines, capable of operating without direct human control. These vehicles navigate through the water with pre-programmed missions, utilizing onboard sensors and instruments to collect data in challenging marine environments. AUVs must have the capability to detect damage, such as leaks, and accurately estimate their location relative to the identified damage [6]. AUVs also offer the potential for cost savings by minimizing the need for expensive surface support operations. However, AUVs are limited in adaptability and real-time decision-making in unpredictable underwater environments, relying on extensive and precise pre-mission planning. To overcome such limitations of AUVs, advancements in artificial intelligence, sensory technology, underwater communication, software flexibility, and energy management are essential.

1.2.1.3 Pipeline Crawlers

Pipeline crawlers are robotic devices that attach directly to the pipeline and move along its length, offering close-up visual inspections [7]. They focus on capturing and relaying real-time visual data. This visual data allows for the identification of surface anomalies, corrosion, mechanical damages, and other external defects that can be seen with the naked eye or camera lens. Some crawlers can be equipped with Non-Destructive Testing (NDT) sensors to extend their inspection capabilities. Their autonomous operation reduces human risk and can decrease long-term costs by lessening the need for diver interventions and support vessels. However, Pipeline crawlers face operational challenges, including the need for alternative propulsion methods beyond hydraulic pressure, complicating navigation through pipelines with thick coatings or vertical sections. Inspecting through concrete and protective layers may require costly pipeline modifications, such as removing marine growth. Structural complexities and environmental factors like underwater currents further limit their effectiveness.

1.2.1.4 Sonar Imaging

Sonar imaging, utilized extensively for underwater navigation and mapping, employs sound waves to detect objects. Sonar imaging uses sound waves in two ways: passive sonar (PS) picks up sounds, while active sonar (AS) emits sound pulses that echo off objects to detect and locate them, based on echolocation principles [8]. The time it takes for the echoes to return is measured. This allows the sonar system to calculate the distance to the object, its size, and its shape, thereby creating an image of the underwater landscape. For checking submarine pipelines, AS is vital. It includes methods like multibeam bathymetry, which maps the sea floor and spots pipeline issues by timing sonar echoes, and side scan sonar (SSS). Newer techniques like wideband sonar (WBS) promise higher-resolution images. Its advantages include the capacity to cover large areas quickly and to detect pipelines and anomalies even in murky waters where other visual inspection methods fail. Active sonar, in particular, can provide detailed images of the pipeline's condition, aiding in the identification of potential issues such as leaks or structural weaknesses. However, sonar imaging can be less effective in highly cluttered environments or where the seafloor's composition absorbs or scatters sound waves, reducing image clarity.

Visual inspection techniques offer significant advantages for submarine pipeline assessments, providing immediate visual data that facilitates quick decision-making for maintenance and repairs. These methods are crucial for verifying the effectiveness of repairs and installations. However, their effectiveness can be limited by factors such as water clarity and insufficient light, especially in deep or turbid waters, often requiring additional measures for effective inspection. Additionally, visual inspections mainly provide surface-level evaluations and may not detect internal or sub-surface issues without further investigation. In summary, while visual inspection methods are vital for assessing submarine pipeline integrity, their limitations necessitate a comprehensive approach that combines various inspection techniques.

1.2.2 Non-destructive Testing Methods

Non-Destructive Testing (NDT) Methods encompass a range of crucial techniques, including ultrasonic testing and radiography, tailored for inspecting both steel and composite submarine pipelines without inflicting any damage [9]. These methods are instrumental in detecting, characterizing, and evaluating defects, ensuring the structural integrity and operational safety of the pipelines. NDT techniques vary in their application, with some being more suited to specific types of materials or defects, thereby highlighting the importance of selecting the appropriate method for a given inspection task. The adoption of NDT methods not only plays a pivotal role in maintaining the longevity and reliability of critical underwater infrastructure but also underscores the commitment to operational excellence and environmental safety in the energy sector.

1.2.2.1 Magnetic Flux Leakage (MFL)

MFL is used to inspect the integrity of ferromagnetic materials like steel pipelines. It operates by magnetizing the pipeline and provides a rapid and efficient means of detecting areas where magnetic fields leak due to corrosion, pitting, or other forms of damage. While MFL offers the advantage of identifying and locating subsurface defects without excavation, its effectiveness is limited by the pipeline's coating thickness, the orientation of the magnetic field, and the presence of external ferrous materials, which can obscure readings [10]. The detection of MFL signifies an anomaly within the pipe wall, necessitating detailed signal analysis to elucidate the nature of the defect. Moreover, MFL may struggle to detect defects in areas with complex geometries or tight bends, potentially leading to missed anomalies. Figure 1.3 shows how the orientation of the magnetic field impacts the defect detection. Advanced analytical and numerical simulations of MFL, such as finite element analysis (FEA) and boundary element method (BEM), have been employed to accurately deduce the geometry of defects from the magnetic field signals measured, enhancing the precision of pipeline inspections [11]. Recent advancements in MFL technology include enhanced sensitivity and data analysis algorithms, improving defect detection accuracy. However, MFL's reliance on magnetic properties makes it unsuitable for inspecting composite pipelines, which are non-ferromagnetic and require alternative inspection methods like ultrasonic testing or thermography.

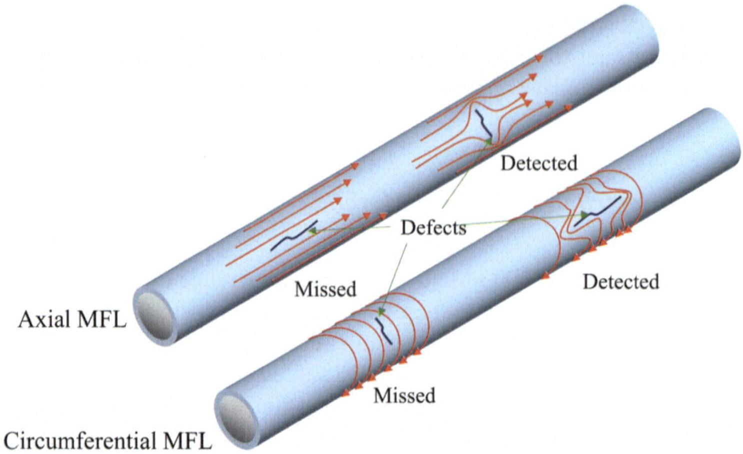

Fig. 1.3 Application of MFL with different orientations of the magnetic field

1.2.2.2 Ultrasonic Testing (UT)

UT is a non-destructive inspection method that uses high-frequency sound waves to detect flaws, measure thickness, and evaluate material properties in pipelines. In pipeline inspection, a transducer emits ultrasonic waves into the pipeline wall; these waves reflect back upon encountering any discontinuities, such as cracks or corrosion, providing detailed information about the location and size of the defect. This technique is highly valued for its ability to accurately assess the integrity of both steel and composite pipelines. UT is non-invasive, requiring no direct access to the entire pipeline, thus suiting underwater applications with limited accessibility. It's safe, free from ionizing radiation, and allows for remote operation, significantly enhancing inspection safety. However, UT also has its limitations. The quality of the data can be affected by the surface condition of the pipeline, with rough, irregular, or coated surfaces potentially leading to less reliable inspections. Additionally, for composite materials, the heterogeneous nature can complicate the propagation of ultrasonic waves, requiring specialized equipment and techniques to achieve accurate results. The effectiveness of UT in composite pipelines can also be limited by the material's acoustic properties, which may dampen or scatter the ultrasonic signals.

Guided Wave Testing (GWT): GWT employs low-frequency ultrasonic waves for non-destructive inspection of steel and composite submarine pipelines, enabling the detection of defects over long distances from a single point [12]. This method excels in efficiently scanning extensive pipeline sections, including those that are inaccessible, coated, or buried, by reflecting waves off irregularities like corrosion or cracks to pinpoint their locations. While GWT offers the advantage of broad coverage and cost-effectiveness, its sensitivity to small defects may be lower compared to other techniques, and it demands expert interpretation of data.

1.2.2.3 Radiography

Radiography uses electromagnetic radiation, such as X-rays and gamma rays, to produce images of a pipeline's interior, revealing defects such as corrosion, cracks, and weld flaws in both steel and composite submarine pipelines [13]. The method works by directing radiographic waves through the pipeline; variations in material density or thickness alter the intensity of the waves that reach the detector, creating a detailed image of the internal structure and any existing anomalies. Radiography inspection is highly beneficial for inspecting submarine pipelines made of both steel and composite materials, thanks to its capability to penetrate thick walls, deliver high-resolution images, and yield quantitative data on defects. However, it faces challenges including radiation hazards, the need for thorough surface preparation, and access limitations, particularly for steel pipelines. In the case of composite pipelines, although radiography is non-destructive and effective in identifying issues like delamination and voids, its performance may be affected by limitations in penetration depth and resolution, along with the complexity of data interpretation, potentially impacting its overall effectiveness. Moreover, for onshore pipelines, personnel can be directly dispatched to conduct radiography; however, for submarine pipelines, the necessary radiography equipment can be delivered using robotic vehicles.

1.2.2.4 Eddy Current Testing (ECT)

ECT uses electromagnetic induction to detect surface and near-surface defects in conductive materials. In submarine pipeline inspection, a probe containing a coil is passed over the pipeline's surface; when an alternating current flows through the coil, it generates a magnetic field that induces eddy currents in the conductive material of the pipeline. Variations in the flow of these eddy currents, caused by defects or material changes, are detected by the probe and used to identify potential issues without making physical contact with the pipeline, as shown in Fig. 1.4. ECT is advantageous for submarine pipeline inspections, enabling non-contact detection of surface and near-surface defects and offering rapid scanning with immediate results, ideal for metallic pipelines [14]. However, its effectiveness is limited in non-conductive materials like composites and in detecting deeper subsurface flaws. Additionally, underwater marine growth and coatings can affect ECT's accuracy, often requiring prior cleaning. Enhanced sensor technologies and array probes have improved the detection and characterization of defects, while innovations in signal processing and data analysis have refined result interpretation, minimizing false positives. The integration of ECT with robotics has facilitated inspections in challenging environments, such as underwater or confined spaces, streamlining the inspection process.

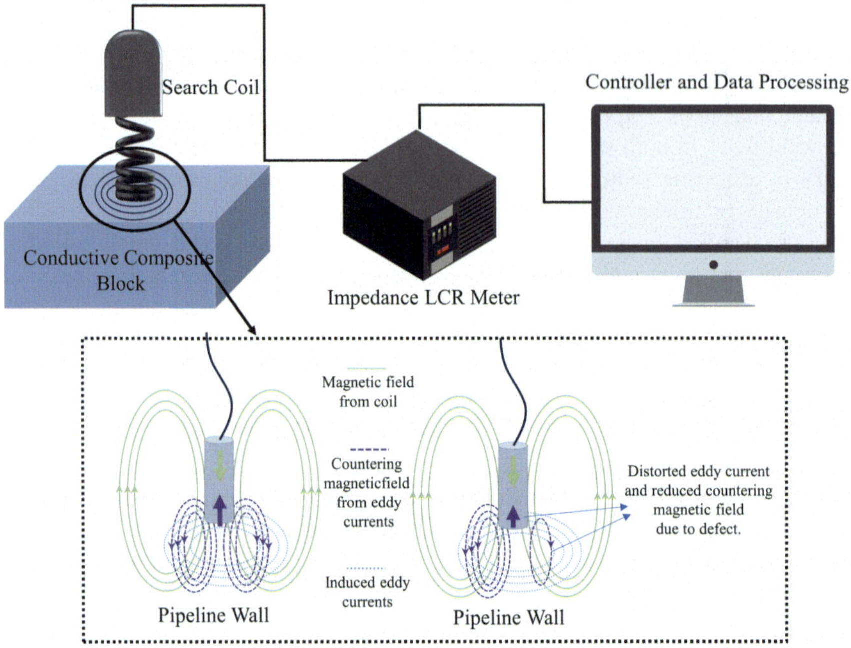

Fig. 1.4 The working setup and principle of eddy current testing

1.2.2.5 Thermographic Inspection

Thermographic Inspection, or thermal imaging, utilizes infrared radiation to detect defects in materials, captured via an infrared camera. This technique identifies changes in thermal behavior due to defects like impact damage or delamination in steel and composite submarine pipelines, revealing localized differences in infrared emission. Defects closer to the surface alter thermal emissions more significantly than those deeper within, making them more detectable by this method. Thermographic inspection offers the advantage of inspecting large surface areas without direct contact, making it ideal for examining pipelines where only one side is accessible. It can be conducted in passive mode, measuring surface temperatures to identify hot spots indicating potential issues, or in active mode, where thermal excitation highlights defects through abnormal heat transfer responses [15]. This method is particularly effective for the NDE of composite parts and can detect a range of issues including delamination, cracks, and water ingress. However, its effectiveness diminishes for defects located deep below the surface, and it requires sensitive, expensive equipment and skilled operators. The lack of physical coupling allows for versatile inspections, but the depth and clarity of detectable defects are limited, necessitating a balanced approach to its application in pipeline inspections.

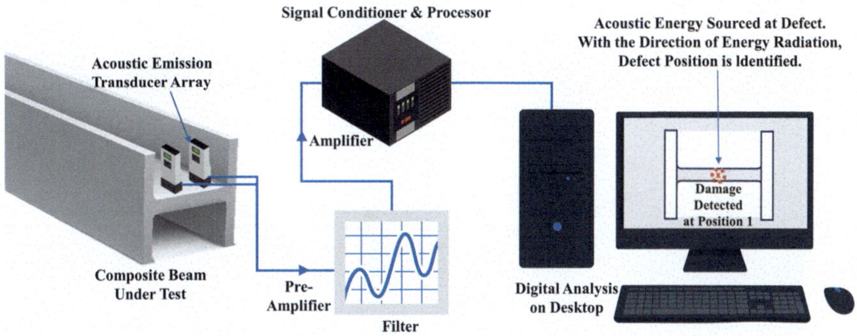

Fig. 1.5 The working setup of acoustic emission testing

1.2.2.6 Acoustic Emission Testing (AET)

AET leverages the stress waves generated by defects such as localized delamination, fiber fracture, and matrix cracking within materials [16]. These stress waves propagate radially from their source and are detected by highly sensitive piezoelectric sensors. Unique to AET is its reliance on the inherent release of sound energy from within the material under stress, rather than external energy application, enabling it to identify dynamic faults as they develop. AET stands out for its rapid, comprehensive testing capabilities using multiple sensors and its high sensitivity, allowing for real-time monitoring of defect evolution under fatigue loading, including fiber/matrix debonding, fatigue cracks, and delamination. This method is particularly adept at detecting a range of defect sizes, from sub-millimeter to 5 mm. However, its effectiveness hinges on the accurate interpretation of acoustic data to specific damage mechanisms, a process that demands a high level of expertise. Additionally, acoustic signals can undergo significant alterations as they travel through composite structures, complicating their interpretation. A typical working setup of acoustic emission testing is shown in Fig. 1.5.

1.2.2.7 Shearography

Shearography, a laser-based non-destructive testing method, illuminates a sample with a laser and captures deformation through a CCD camera, creating a shearogram by superimposing laterally shifted images, as shown in Fig. 1.6. This technique is adept at detecting surface and subsurface defects in composite materials by observing disturbances in the interference fringe pattern under various loading conditions. It is highly effective in identifying delamination, cracks, and other defects by highlighting stress concentrations, making it particularly useful for inspecting composite submarine pipelines. Shearography is not suitable for steel pipelines primarily because it relies on detecting surface deformations caused by subsurface defects, which is less effective on the highly conductive and less deformable surfaces of metallic

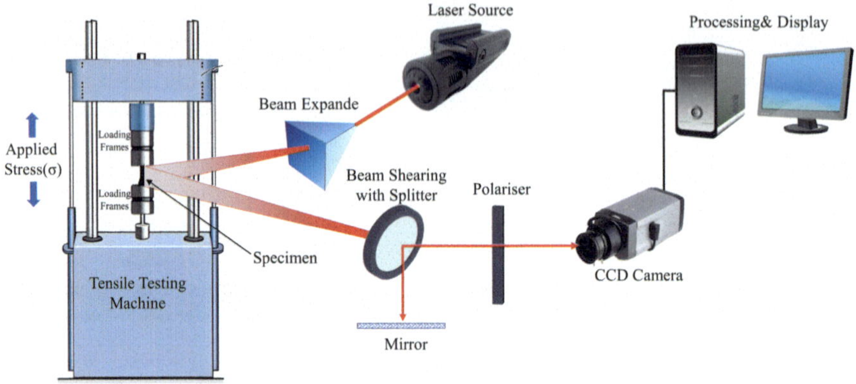

Fig. 1.6 The working setup of shearography testing

materials [17]. Shearography offers significant advantages for inspecting composite materials, including the ability to detect defects up to 2.6 mm in depth without direct contact, making it ideal for honeycomb and foam structures. Its sensitivity to stress concentrations around defects allows for the identification of the type and criticality of flaws, with less susceptibility to noise compared to other NDE methods. However, shearography's effectiveness diminishes with the depth or diameter increase of the defect, requiring appropriate external loading systems for examination. This method's primary limitation is its difficulty detecting non-delamination defects, often necessitating a combination with other NDE techniques for a comprehensive defect characterization. The method requires external loading and involves the complexity of digital shearography applications. These applications depend on various factors such as defect depth, type, material type, and laser intensity, which present challenges for widespread adoption.

1.2.2.8 Fiber Optic Sensor (FOS)

FOS is an advanced sensor technology utilizing strands of silica glass to transmit light for monitoring both onshore and offshore pipelines. These sensors, originally developed for communications, exploit properties of light such as intensity, wavelength, phase, and polarity, modulated by environmental changes to detect anomalies [18]. In submarine pipeline monitoring, FOS can be either point sensors, offering high resolution at single points, or distributed sensors, using the entire fiber as a continuous sensor array, though with lower resolution but broader coverage. Fiber Bragg Grating (FBG) sensors, a type of quasi-distributed FOS, are particularly noted for their ability to detect mechanical changes like bending and impact through wavelength modulation. FBGs are inscribed with gratings that reflect specific wavelengths altered by strain, temperature, and pressure, making them suitable for real-time monitoring of subsea pipelines. These sensors can be encapsulated and retrofitted onto existing pipelines, allowing for non-intrusive detection of conditions affecting flow assurance, such as deposit buildup or leakage. Real-time monitoring with FOS, especially FBGs, enables early remedial actions, potentially avoiding costly pipeline replacements. However, the complexity of deploying and interpreting data from FOS, along with their sensitivity to environmental conditions, still poses challenges.

1.3 Hybrid Inspection Methods

The subsea environment presents a highly intricate and dynamic. Coupled with the inherent maintenance demands of pipelines, the process of submarine pipeline inspection and monitoring becomes a challenging endeavor. It necessitates the effective use of multiple tools concurrently to attain optimal outcomes. Combining visual, NDT, and smart technologies offers a complete strategy for inspecting submarine pipelines. This method uses each technique's strengths to complement the limitations of others. For example, visual checks, NDT methods like ultrasonic tests and radiography, and smart tools such as ROVs with sensors, provide a detailed view of the pipeline's state from the surface to deep inside. This mix improves how well defects are found by giving different views on the pipeline's health, making it easier to spot and describe issues accurately.

Using all these methods together to only does it improve defect detection but also helps manage the pipeline's condition more effectively. It allows for a smarter way to plan maintenance and repairs, which lowers the chance of downtime and failures. This approach leads to safer operations, better efficiency, and less environmental impact, showing the value of integrating various inspection methods for the best results in managing pipeline integrity.

Tables 1.1 and 1.2 list the inspection methods and their advantages and disadvantages for steel and composite submarine pipelines, respectively.

1.4 Future Directions in Pipeline Inspection

Inspecting submarine pipelines presents several challenges, including accessing pipelines buried deep underwater with high turbulence, effectively detecting and characterizing a range of defects, and limitations in current technologies' ability to provide comprehensive condition assessments. The use of different materials, like steel and composites, in pipeline construction increases the complexity, as each material has unique inspection needs. The marine environment further complicates inspections due to its corrosive nature and the need for technologies capable of withstanding high pressures and extreme temperatures.

In contrast, innovations in signal processing and neural networks enhance defect detection, characterization, and complex damage identification. Development of Frequency Modulated Continuous Wave (FMCW) radar sensing with digital twin

Table 1.1 NDT inspection methods for steel submarine pipelines

NDT method	Detectable defects/ flaws	Advantages	Disadvantages
Magnetic flux leakage (MFL)	Cracks, corrosion, metal loss, weld defects, pitting, liner defects, gouging	High-speed inspection, can inspect through coatings, quantitative data	Limited to ferromagnetic materials, less effective for complex geometries
Ultrasonic testing (UT)	Cracks, thickness variations, corrosion, weld defects	High depth penetration, precise defect sizing, versatile for various materials	Requires skilled operator, surface condition affects accuracy
Radiography	Cracks, corrosion, weld defects, material density changes, voids, slag inclusions	High-resolution images, deep penetration, can inspect complex geometries	Health hazard due to radiation, access limitations for thick materials
Eddy current testing (ECT)	Surface cracks, coating defects, small metal loss, near-surface defects	Sensitive to small cracks and surface defects, no need for direct contact	Limited to conductive materials, depth penetration is shallow
Thermographic INSPECTION	Coating defects, corrosion under insulation, leaks, voids	Non-contact, large areas quickly inspected, detects subsurface defects	Affected by environmental conditions, requires experienced operator
Acoustic Emission Testing (AET)	Crack propagation, active corrosion, leakage	Detects active defects, covers large areas, suitable for real-time monitoring	Background noise can affect accuracy, complex data interpretation
Fiber optic sensors	Temperature changes, strain, vibration, bending, cracks, corrosion, erosion, physical damage, leakage	High sensitivity, immune to electromagnetic interference, real-time monitoring	Installation can be complex, initial cost is high

technologies leaps forward into improved defect detection and integrity management of pipelines in real-time. These technologies offer non-contact, non-destructive testing capabilities that are unaffected by environmental conditions, providing a versatile and efficient solution to the challenges of subsea pipeline inspection. On the other hand, the direction is increasingly going toward an integrated approach to inspecting submarine pipelines using both traditional and novel NDE methods by digital analysis and AI. This holistic approach, leveraging the best of different sensing technologies and data analytics, will ensure the continued safety and reliability of submarine pipelines—part of the backbone of the global energy infrastructure.

Table 1.2 NDT inspection methods for composite submarine pipelines

ndt method	Defects/flaws detected	Advantages	Disadvantages
Ultrasonic testing (UT)	Delamination, voids, disbonds, thickness variations, fiber waviness, porosity	High precision and depth penetration, applicable to complex geometries	Requires coupling medium, skill-dependent interpretation, limited by material acoustical properties
Radiography	Delamination, voids, fiber orientation, foreign objects, density changes	High-resolution imaging, effective for complex defects	Health hazards from radiation, limited access scenarios, high operational costs
Eddy current testing (ECT)	Surface cracks, coating flaws (limited in composites)	Effective for surface and near-surface defects, fast inspection rates	Limited to conductive materials, less effective for deep defects
Thermographic Inspection	Delamination, voids, disbonds, impact damage, water ingress, thickness variations	Non-contact, large areas inspected quickly, detects surface and near-surface defects	Requires controlled conditions, less effective for deep defects, interpretation can be complex
Acoustic emission testing	Fiber breakage, matrix cracking, delamination, fiber/matrix debonding	Can detect active defect propagation, covers large areas, sensitive to various defect types	Complex data analysis, requires quiet operational environment, less effective for initial defect detection
Shearography	Delamination, disbonds, voids, impact damage, water ingress, structural integrity under load	Non-contact, sensitive to surface and subsurface defects, can inspect large areas quickly	Requires specific loading conditions, limited depth penetration, complex setup
Fiber optic sensors	Strain, temperature changes, vibration, acoustic signals, cracks growing, delamination, impact Damage, leaks	Real-time monitoring, immune to electromagnetic interference, can cover long distances	Installation can be complex, high initial cost, interpretation requires specialized knowledge

References

1. Du F, Li C, Wang W (2023) Development of Subsea Pipeline Buckling, Corrosion and Leakage Monitoring. Journal of Marine Science and Engineering 11(1):188. https://doi.org/10.3390/jmse11010188
2. Patel S, Abdellatif F, Alsheikh M, Trigui H, Outa A, Amer A, Sarraj M, Al Brahim A, Alnumay Y, Felemban A, Alrasheed A, Halawani A, Jifri H, Jaleel H, Shamma J (2024) Multi-robot system for inspection of underwater pipelines in shallow waters. Int J Intell Robot Appl. https://doi.org/10.1007/s41315-023-00309-8

3. Fun Sang Cepeda M, de Souza Freitas Machado M, Sousa Barbosa FH, Santana Souza Moreira D, Legaz Almansa MJ, de Souza MIL, Caprace J-D (2023) Exploring autonomous and remotely operated vehicles in offshore structure inspections. J Mar Sci Eng 11(11):2172.https://doi.org/10.3390/jmse11112172
4. Spahić R, Poolla K, Hepsø V, Lundteigen MA (2023) Image-based and risk-informed detection of Subsea Pipeline damage. Discov Artif Intell 3(1):23. https://doi.org/10.1007/s44163-023-00069-1
5. Ierodiaconou D, McLean D, Birt MJ, Bond T, Wines S, Glade-Wright O, Morris J, Higgs D, Whitmarsh SK (2023) Industry remotely operated vehicle imagery for assessing marine communities associated with subsea oil and gas infrastructure on the continental shelf of South-East Australia. Front Mar Sci 10.https://doi.org/10.3389/fmars.2023.1095906
6. Feng H, Yu J, Huang Y, Cui J, Qiao J, Wang Z, Xie Z, Ren K (2023) Automatic tracking method for submarine cables and pipelines of AUV based on side scan sonar. Ocean Eng 280:114689. https://doi.org/10.1016/j.oceaneng.2023.114689
7. Vagata A (2023) Implementation of a multi-technology approach for flexible risers' inspection—improving effectiveness and efficiency. Paper presented at the offshore technology conference Brasil, Rio de Janeiro, Brazilhttps://doi.org/10.4043/32925-MS
8. Shi B, Cao T, Ge Q, Lin Y, Wang Z (2024) Sonar image intelligent processing in seabed pipeline detection: review and application. Meas Sci Technol 35(4):045405. https://doi.org/10.1088/1361-6501/ad1919
9. Waqar M, Memon AM, Sabih M, Alhems LM (2024) Composite pipelines: analyzing defects and advancements in non-destructive testing techniques. Eng Fail Anal 157:107914. https://doi.org/10.1016/j.engfailanal.2023.107914
10. Wu D, Liu Z, Wang X, Su L (2017) Composite magnetic flux leakage detection method for pipelines using alternating magnetic field excitation. NDT E Int 91:148–155. https://doi.org/10.1016/j.ndteint.2017.07.002
11. Augustyniak M, Usarek Z (2016) Finite element method applied in electromagnetic NDTE: a review. J Nondestruct Eval 35(3):39. https://doi.org/10.1007/s10921-016-0356-6
12. Ghavamian A, Mustapha F, Baharudin BTHT, Yidris N (2018) Detection, localisation and assessment of defects in pipes using guided wave techniques: a review. Sensors 18(12):4470. https://doi.org/10.3390/s18124470
13. Li X-F, Sun J, Lu S, Wang L (2022) Application of on-line digital radiographic inspection for pipeline with insulation. J Phys: Conf Ser 2366(1):012006. https://doi.org/10.1088/1742-6596/2366/1/012006
14. AbdAlla AN, Faraj MA, Samsuri F, Rifai D, Ali K, Al-Douri Y (2019) Challenges in improving the performance of eddy current testing: review. Measur Control 52(1–2):46–64. https://doi.org/10.1177/0020294018801382
15. Cadelano G, Bortolin A, Ferrarini G, Molinas B, Giantin D, Zonta P, Bison P (2016) Corrosion detection in pipelines using infrared thermography: experiments and data processing methods. J Nondestruct Eval 35(3):49. https://doi.org/10.1007/s10921-016-0365-5
16. Ho M, El-Borgi S, Patil D, Song G (2020) Inspection and monitoring systems subsea pipelines: a review paper. Struct Health Monit 19(2):606–645. https://doi.org/10.1177/1475921719837718
17. Farshi SS, Akbari D (2023) Integration of digital shearography and FEM for non-destructive evaluation of internal cracks in GRE pipes. J Reinf Plast Compos 07316844231219302. https://doi.org/10.1177/07316844231219302
18. Ren L, Jiang T, Jia Z, Li D, Yuan C, Li H (2018) Pipeline corrosion and leakage monitoring based on the distributed optical fiber sensing technology. Measurement 122:57–65. https://doi.org/10.1016/j.measurement.2018.03.018

Chapter 2
Corroded Pipeline Assessment

2.1 Introduction

According to the report, around 47% of the damage to offshore steel pipelines in China is caused by corrosion defects, as shown in Fig. 2.1 [1]. In this chapter, the critical need for accurate reliability assessment of corroded pipelines, essential for transporting energy resources, is addressed. Despite extensive research, the interaction between adjacent corrosion defects and the effects of corrosion depth and steel grade on pipeline integrity have been inadequately explored. A novel reliability assessment method is introduced, integrating finite-element analysis, regression, burst pressure models, Monte Carlo simulation, and artificial neural networks to fill this gap. This comprehensive approach refines interaction rules considering corrosion characteristics and enhances prediction efficiency and accuracy. By applying this method to a case study, advancements in pipeline safety management are pursued, aiming to establish a new benchmark for the industry's reliability standards.

2.2 Causes of Pipeline Corrosion

Pipeline corrosion, a pervasive issue affecting both onshore and offshore pipelines, is primarily caused by environmental interactions and the inherent characteristics of the materials used. Onshore pipelines are often subjected to corrosion due to soil characteristics, microbial activity, and atmospheric conditions [2]. The presence of moisture, varying pH levels, and oxygen in the soil can accelerate the corrosion process. Microbial-induced corrosion (MIC) also plays a significant role, as bacteria can thrive in certain soil types, producing acids that deteriorate the pipeline material. Atmospheric corrosion is influenced by pollutants and climatic conditions, leading to rust and pitting. Offshore pipelines, on the other hand, face corrosion challenges from seawater, which is highly conductive and facilitates electrochemical reactions

H. Zhang et al., *Submarine Pipeline Integrity: Assessment of Failure Modes and Advanced Evaluation Techniques*, Ocean Engineering & Oceanography 28,
https://doi.org/10.1007/978-3-031-92092-9_2

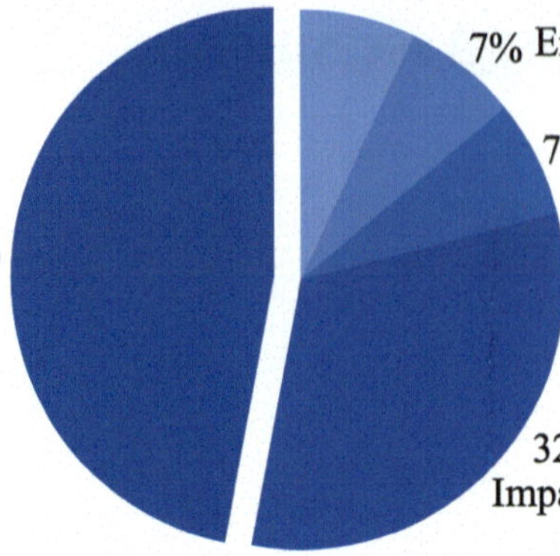

Fig. 2.1 Causes of pipeline damage in China

[3]. The presence of chloride ions, high salinity, and varying temperatures contribute to the severity of corrosion. Additionally, marine organisms can attach to pipelines, causing biofouling that leads to under deposit corrosion. The movement of water and sediments can cause mechanical wear that removes protective coatings, exposing the metal to corrosive elements.

2.3 Burst Models and Interaction Rules

Over the past years, various burst pressure models considering geometric features and material properties have been developed, such as ASME B31G [4], RSTRENG [5], CSA Z662-15 [6], DNV-RP-F101 [7], and PCORRC [8, 9]. Each burst model has its advantages and limitations. Thus, choosing a burst pressure model for the burst pressure assessment should be based on specific conditions.

Developed from the NG-18 model, ASME B31G is one of the most widely used burst models in the energy industry. It incorporates simplified geometric features, which makes it more convenient to use. The ASME B31G burst model is defined as follows [10]:

$$P_b = \begin{cases} \frac{2t}{D}(1.1\sigma_y)\left[\frac{1-\frac{2}{3}\left(\frac{d}{t}\right)}{1-\frac{2}{3}\left(\frac{d}{t}\right)M^{-1}}\right], & \frac{l^2}{Dt} \le 20 \\ \frac{2t}{D}(1.1\sigma_y)\left[1-\frac{d}{t}\right], & \frac{l^2}{Dt} > 20 \end{cases} \tag{2.1}$$

$$M = \sqrt{1 + 0.8\left(\frac{l^2}{Dt}\right)} \tag{2.2}$$

where t and D is the wall thickness and the diameter of the pipeline respectively, σ_y is the yield stress of the pipe steel, d is the corrosion's depth, M is the Folias Factor and l is the length of the corrosion.

To pursue a more accurate estimation of the burst pressure, the ASME B31G burst model was improved by two modifications, i.e., the controlling flow stress and shape-related factors. The expression of the ASME B31G Modified burst model is defined as follows [11]:

$$P_b = \frac{2t}{D}(\sigma_y + 69)\left[\frac{1 - 0.85\left(\frac{d}{t}\right)}{1 - 0.85\left(\frac{d}{t}\right)M^{-1}}\right] \tag{2.3}$$

$$M = \begin{cases} \sqrt{1 + 0.6275\left(\frac{l^2}{Dt}\right) - 0.003375\left(\frac{l^2}{Dt}\right)^2}, & \frac{l^2}{Dt} \le 50 \\ 3.3 + 0.032\left(\frac{l^2}{Dt}\right), & \frac{l^2}{Dt} > 50 \end{cases} \tag{2.4}$$

The CSA Z662-07 burst model divides the pipelines into different types based on the yield stress of the pipe steel, i.e., high-grade steel ($\sigma_y > 241$ MPa) and low-grade steel ($\sigma_y \le 241$ MPa). It is used to describe the plastic collapse at a surface corrosion defect. The CSA Z662-07 model is defined as follows [11]:

$$P_b = \begin{cases} e_1 P_{bi} + (1 - e_1)P - e_2\sigma_u, & \sigma_y > 241\,\text{MPa} \\ e_3 P_{bi} + (1 - e_3)P - e_4\sigma_y, & \sigma_y \le 241\,\text{MPa} \end{cases} \tag{2.5}$$

$$P_{bi} = P\left[\frac{1 - \left(\frac{d_a}{t}\right)}{1 - \left(\frac{d_a}{t}\right)M^{-1}}\right] \tag{2.6}$$

$$P = \begin{cases} \frac{2t(0.9\sigma_u)}{D}, & \sigma_y > 241\text{ MPa} \\ \frac{2t(1.15\sigma_y)}{D}, & \sigma_y \le 241\text{ MPa} \end{cases} \tag{2.7}$$

$$M = \begin{cases} \sqrt{1 + 0.6275\left(\frac{l^2}{Dt}\right) - 0.003375\left(\frac{l^2}{Dt}\right)^2}, & \frac{l^2}{Dt} \le 50 \\ 3.3 + 0.032\left(\frac{l^2}{Dt}\right), & \frac{l^2}{Dt} > 50 \end{cases} \tag{2.8}$$

where e_1 is the deterministic model error, $e_1 = 1.04$, e_2 is an additive model error, $e_2 \sim N\left(-0.00056, 0.001469^2\right)$, e_3 is a deterministic multiplicative model error, $e_3 = 1.17$, e_4 is an additive model error, $e_4 \sim N\left(-0.007655, 0.006506^2\right)$, σ_u is the ultimate tensile strength of the pipe steel, P_{bi} is the calculated pressure strength, P is the pressure strength of the intact pipeline.

Based on the experimental and finite element simulation data, DET NORSKE VERITAS collaborated with BP Technology proposed the DNV RP F-101 burst model [7]. In this model, plastic collapse failure determines the corroded pipeline's defeat. The expression of the DNV RP F-101 can be defined as Eq. 2.9, and the 1.05 in the equation is generated by laboratory test comparison.

$$P_b = 1.05\frac{2t\sigma_u\left(1 - \frac{d}{t}\right)}{(D - t)\left(1 - \frac{d}{t}M^{-1}\right)} \tag{2.9}$$

$$M = \sqrt{1 + 0.31\left(\frac{l^2}{Dt}\right)} \tag{2.10}$$

The PCORRC burst model was proposed by Leis [8] by analyzing the simulation results from a set of finite element analyses. This model aims to estimate the pipeline's ultimate bearing capacity with blunt corrosion defects. The equation of the PCORRC model can be expressed as Eq. 2.11. The function in the bracket of Eq. 2.11 is defined as a Folias Factor. r is the radius of the pipeline.

$$P_b = \frac{2t\sigma_u}{D}\left\{1 - \frac{d}{t}\left[1 - \exp\left(-0.157\frac{l}{\sqrt{r(t-d)}}\right)\right]\right\} \tag{2.11}$$

The SHELL92 burst model was developed based on the ASME B31G model, and modifications were implemented to introduce a certain degree of conservativeness to burst pressure assessment. The SHELL92 burst model is defined as follows [12]:

$$P_b = \frac{2t(0.9\sigma_u)}{D}\left[\frac{1 - \left(\frac{d}{t}\right)}{1 - \left(\frac{d}{t}\right)M^{-1}}\right] \tag{2.12}$$

$$M = \sqrt{1 + 0.8\left(\frac{l^2}{Dt}\right)} \tag{2.13}$$

The RPA burst model was also developed from the ASME B31G model by introducing a higher level of conservativeness to estimate the pipeline's residual strength with long corrosion defects. This model classifies corrosion defects into two different types, i.e., long corrosion $\left(\frac{l^2}{Dt} > 20\right)$ and short corrosion $\left(\frac{l^2}{Dt} \leq 20\right)$. The expression of the RPA burst model is defined as Eqs. 2.14–2.16 [13].

$$P_b = \frac{2t}{D}(\sigma_y + 69)\left[\frac{1 - a\left(\frac{d}{t}\right)}{1 - a\left(\frac{d}{t}\right)M^{-1}}\right] \tag{2.14}$$

$$M = \begin{cases} \sqrt{1 + 0.6275\left(\frac{l^2}{Dt}\right) - 0.003375\left(\frac{l^2}{Dt}\right)^2}, & \frac{l^2}{Dt} \leq 20 \\ 2.1 + 0.7\left(\frac{l^2}{Dt}\right), & \frac{l^2}{Dt} > 20 \end{cases} \tag{2.15}$$

$$a = \begin{cases} 0.85, & \frac{l^2}{Dt} \leq 20 \\ 1 - 0.15\left(64 \times 10^6\right)\left(\frac{l^2}{Dt}\right)^6, & \frac{l^2}{Dt} > 20 \end{cases} \tag{2.16}$$

Table 2.1 Interaction rules

Interaction rule	Longitudinal limit (mm)
DNV-RP-F101	$S_{lim} = 2\sqrt{Dt}$
Coulson and Worthingham	$S_{lim} = \min(l_1, l_2)$
ASME B31G	$S_{lim} = 3t$
Kiefner and Vieth	$S_{lim} = \min(6t, l_m, l_n)$

Because of assumptions of the yield criteria and the corrosion geometric features, the PCORRC model might introduce some level of conservatism. The calibration process expressed in Eq. 2.17 should be considered to improve prediction accuracy. Detailed information about Eq. 2.17 can be found in the study conducted by Bhardwaj et al. [14].

$$P^{CB}_{corr} = X^{G}_{corr} \times P_b \tag{2.17}$$

where X^{G}_{corr} is the global uncertainty factor.

For a pipeline with multiple corrosion defects, the volumetric method is usually combined with burst models to quantify the interacting effect. However, not every adjacent corrosion will interact with others. Therefore, the limit spacing distance is commonly used as the threshold to determine whether the interacting effect should be considered. Over the past decades, different companies and researchers have developed several interaction rules. Some typical interaction rules are listed in Table 2.1. However, almost no existing interaction rules consider the effects of the corrosion depth and pipe steel grade. Such ignorance might induce inaccuracy in the burst pressure assessment or further reliability analysis.

2.4 Failure Criteria for Corroded Pipeline

Many failure criteria are available for determining the failure of a corroded pipeline, such as the 80% of the wall thickness failure criterion, the plastic failure criterion, the yield failure criterion, and so on. Among the existing failure criteria, the Von Mises yield criterion and limit state function are the most widely used.

The Von Mises yield criterion determines if a corroded pipeline will suffer from plastic collapse. Once the Von Mises stress (σ_v) exceeds the stress limit, and a failure occurs. According to previous research, setting the ultimate tensile strength (UTS) as the stress limit could improve prediction accuracy. The Von Mises yield criterion, expressed by Eq. 2.18, is employed in the finite-element analysis.

$$\sigma_v \leq \text{UTS} \tag{2.18}$$

However, the Von Mises yield criterion is no longer useful in the reliability analysis. According to the structure reliability theory, the LSF, expressed as Eq. 2.19, is used to determine a failure.

$$g = P_{corr}^{CB} - P \tag{2.19}$$

where P_{corr}^{CB} is the calibrated burst pressure, P is the internal pressure. If the operating pressure is within the calibrated burst pressure, i.e., $g > 0$, the corroded pipeline is in a safe operating state; otherwise, the corroded pipeline has failed.

2.5 New Interaction Rule

FEM is applied to investigate how the corrosion depth and steel grade influence the limit spacing distance. Performing full-scale experiments could be expensive and challenging to keep the experiment results satisfactory. Meanwhile, FEM has many remarkable merits, such as relatively low cost and high efficiency. Using FEM to address nonlinear problems about corroded pipelines has been proven to be cost-effective and could obtain satisfactory results. Based on the simulation results and regression analysis, new interaction rules considering the effects of the corrosion depth and pipe steel grade can be developed.

2.5.1 *Numerical Simulation Model*

A quarter-symmetry 3D FE model of a pipeline with axial corrosion, as depicted in Fig. 2.2, was efficiently simulated in Abaqus 6.14, utilizing geometric and loading symmetries to balance computational efficiency with accuracy. The model's dimensions, sourced from study [15], featured an outer diameter of 458.8 mm, a wall thickness of 10 mm, and a length of 1500 mm, with corrosion defects exclusively on the outer wall, each 52 mm long and 32 mm wide. Prior findings suggest corrosion width minimally impacts interaction behaviors. The model utilized eight-node hexahedron elements, with a refined 1 mm mesh around defects and a coarser 3 mm mesh elsewhere, a meshing strategy validated for reliability. The analysis included X65 and X80 pipeline steels, whose properties are detailed in Table 2.2. To account for the influence of steel hardening on the pipeline's mechanical response, the true stress–strain relationships for X65 and X80 steels were incorporated, as illustrated in Fig. 2.3, referencing studies [16, 17].

The validity and reliability of finite-element simulations for corroded pipeline failures are confirmed by comparing simulation outcomes with established experimental data [16, 17]. Key parameters, geometric details, and corrosion defect distributions from these experiments are summarized in Table 2.3.

Figure 2.4 shows the burst pressure calculated by FEM and the experimental burst pressure of each testing specimen mentioned previously. The four cases' average relative error is only 3.7%, which is acceptable when conducting FEM. Therefore, the

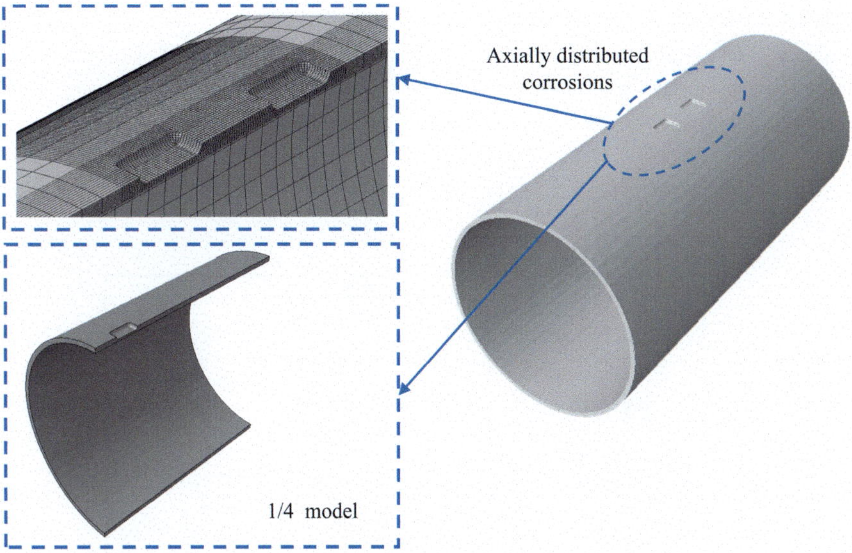

Fig. 2.2 Three-dimensional finite-element model of the pipeline with axially distributed corrosion defects

Table 2.2 Material properties of X65 and X80 steel

Property	X65 steel [18]	X80 steel
Young's modulus (GPa)	206	200
Yield strength (MPa)	456	534.1
UTS (MPa)	565	718.2

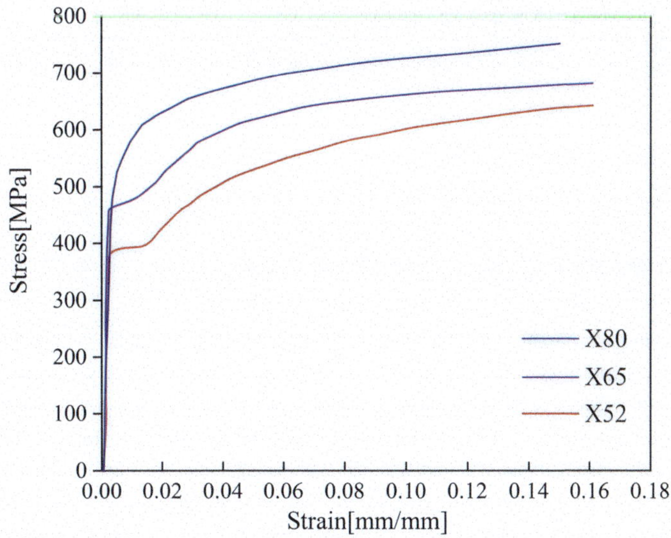

Fig. 2.3 True stress–strain curves of pipeline steels

Table 2.3 Parameters of experimental pipe specimens

Specimen	Steel	D (mm)	t (mm)	d (mm)	l (mm)	w (mm)	l_c (mm)	l_c (mm)	Distribution
IDTS-2	X80	458.8	8.1	5.39	39.6	21.9	–	–	Isolated
IDTS-3	X80	458.8	8.1	5.32	39.6	31.9	20.5	–	Axial
IDTS-4	X80	458.8	8.1	5.62	39.6	32	–	9.9	Circumferential
X65-DB	X65	762	17.5	8.8	200	50	–	–	Isolated

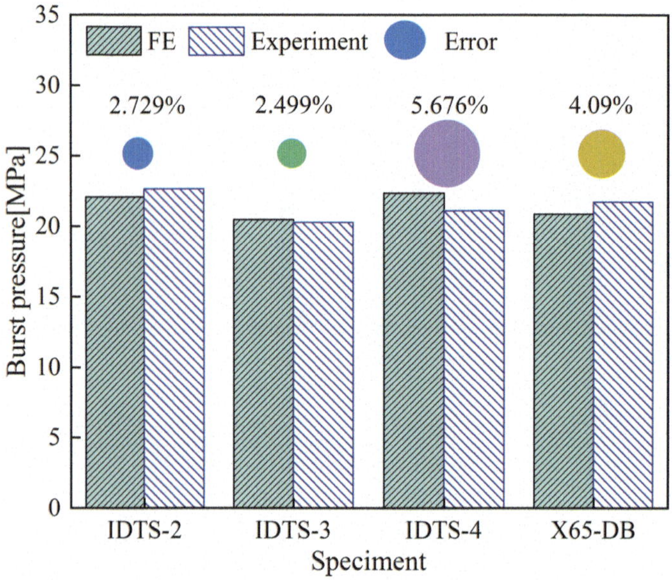

Fig. 2.4 Comparison of burst pressures obtained by experiments and FE simulations

numerical simulation models were validated to be feasible and reliable in assessing the corroded pipeline's burst pressure.

2.5.2 Simulation Results and Discussion

The corrosion geometric features were our major concern for a specific pipeline, while other parameters, like pipeline wall thickness, were already determined. To investigate the effects of corrosion geometric features, including depth-to-thickness ratio (DTR), corrosion length (l), and corrosion width (w), on the burst pressure and interacting effect, an $L_4\left(2^3\right)$ orthogonal test was designed and performed. The testing results, which were analyzed by the range analysis, are listed in Table 2.4. Range value can quantify the test index's variation caused by the variation in input.

Figure 2.5 shows the range values of d, l, and w. It is apparent that DTR has the most significant contribution to burst pressure and interacting effect.

About 120 numerical simulation models with different geometric features were built and simulated according to the aforementioned parameters. Equation 2.20 is a rule widely used in the research of pipelines with interacting corrosion defects to determine if the longitudinally aligned corrosion defects should be regarded as interacting defects [19].

$$\frac{P_{mul}}{P_{iso}} < 0.99 \tag{2.20}$$

where P_{mul} and P_{iso} = burst pressures of a pipeline with multiple and isolated corrosion defects, respectively.

If the ratio between the burst pressure of the pipeline with multiple corrosion defects and that of the pipeline with single corrosion is smaller than 0.99, the interacting effect between corrosion defects is supposed to be considered.

The effects of the corrosion depth and spacing distance on P_{mul}/P_{iso} of the X65 and X80 pipelines are shown in Fig. 2.6. The value of P_{mul}/P_{iso} becomes much lower

Table 2.4 $L_4(2^3)$ orthogonal test

Specimen	DTR	l (mm)	w (mm)	P_{iso} (MPa)	P_{mul} (MPa)	P_{mul}/P_{iso} (%)
1	0.1	40	20	25.05	25.02	99.86
2	0.4	160	40	24.69	24.68	99.95
3	0.6	40	20	21.57	19.6	89.72
4	0.9	160	40	8.26	8.2	99.25

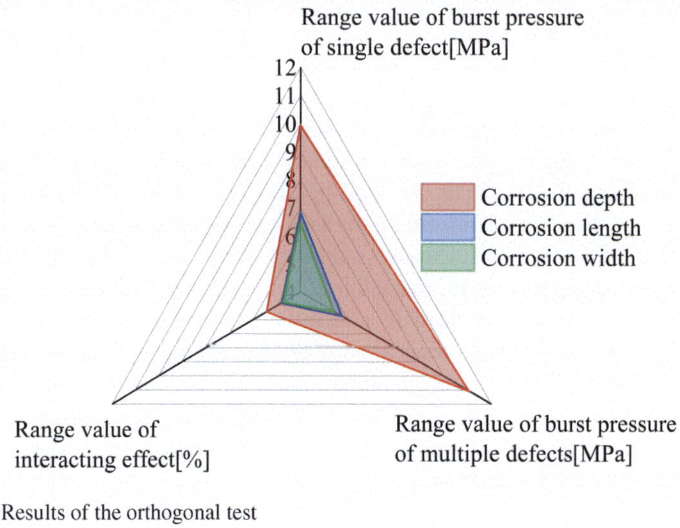

Fig. 2.5 Results of the orthogonal test

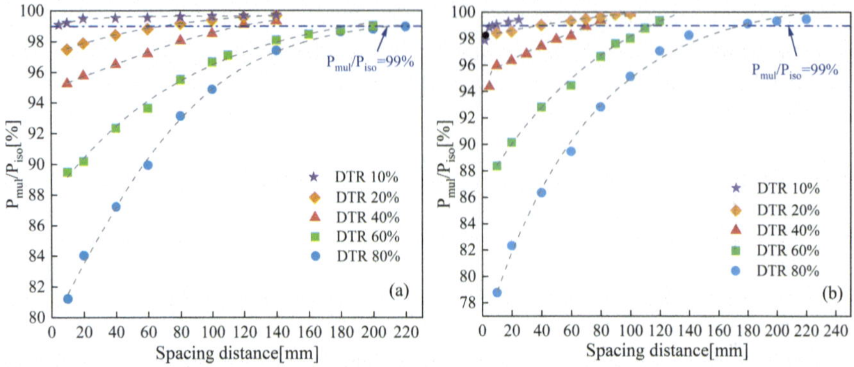

Fig. 2.6 Effects of corrosion depth and spacing distance on P_{mul}/P_{iso} of the corroded pipeline **a** X65, **b** X80

when the corrosion depth grows, which indicates that the extent of the interacting effect between adjacent corrosion defects increases to a higher level once the depth of the corrosion defect grows. The limit spacing distance is defined as the spacing distance that makes $P_{mul}/P_{iso} = 0.99$. Based on the simulation results, the limit spacing distances under different corrosion depths can be inferred.

Based on the simulation results, the limit spacing distances of the X65 and X80 pipelines with different DTRs are shown in Fig. 2.7. The limit spacing distances of both the X65 pipeline and the X80 pipeline are sensitive to DTR variations. With DTR increases, the limit spacing distance shows an upward trend for both the X65 and X80 pipes. The limit spacing distance of the X80 pipe with 80% DTR is about 124 times larger than that with 10% DTR. For the X65 corroded pipeline, when DTR increases from 10 to 80%, there is a 22-fold growth in the limit spacing distance. Thus, the increase in the steel grade could enhance the corrosion depth's effect on the limit spacing distance. When the DTR is larger than 10%, there exists a significant difference between the limit spacing distance of the X65 pipe and that of the X80 pipe at the given DTR.

Despite the existence of several interaction rules, none quantifies the effects of the corrosion depth and the steel grade. When the DTR is 10%, for the X65 and X80 pipes, the limit spacing distances obtained by FEM are smaller than 10 mm, while the limit spacing distances estimated by the four interaction rules are all greater than 30 mm. The limit spacing distance assessed by the DNV interaction rule is 80 times larger than that evaluated by the simulation results of the X80 pipeline. In other words, the DNV interaction rule will induce unnecessary conservatism in further assessment. When the DTR is larger than 40%, K&V, Coulson and Worthingham (CW), and ASME B31G will underestimate the limit spacing distance. Therefore, developing new interaction rules that consider the effects of the corrosion depth and pipeline steel grade is of great value for improving the accuracy of the burst pressure estimation and the reliability analysis.

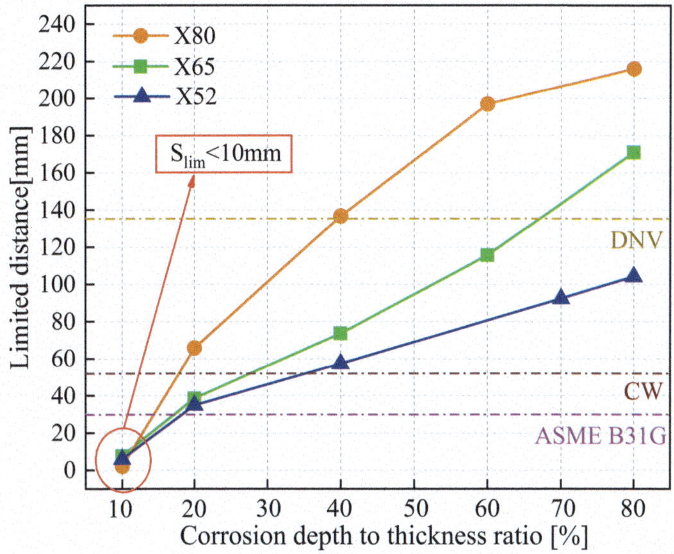

Fig. 2.7 Effect of DTR on the limit spacing distance between adjacent corrosion defects

Figure 2.8 shows the nonlinear relationships between the normalized limit spacing distances and the normalized corrosion depths of the X65 and X80 pipelines. According to the scatter points' regularities, the asymptotic function was employed to fit the limit spacing distances under different normalized corrosion depths. The solid lines in Fig. 2.8 are the fitted lines for the X65 and X80 pipes. The average R^2 value of these two fittings is larger than 0.99, which manifests the goodness of the fit. Through regression analysis, the new interaction rules incorporating the corrosion depth are expressed as Eqs. 2.21 and 2.22 for the X65 pipe and the X80 pipe, respectively.

$$\text{X65}: \quad S_{lim} = \left(3954.875 - 3956.144 \times 0.994^{\frac{d}{t}}\right) \times t \tag{2.21}$$

$$\text{X80}: \quad S_{lim} = \left(27.831 - 33.22 \times 0.132^{\frac{d}{t}}\right) \times t \tag{2.22}$$

2.6 Integrated Reliability Analysis Method for Corroded Pipelines

In most previous relevant studies, the reliability analysis of corroded pipelines was based on isolated corrosion defects. A new reliability method is proposed to fill the gap in the reliability analysis that ignores the interacting effect between adjacent

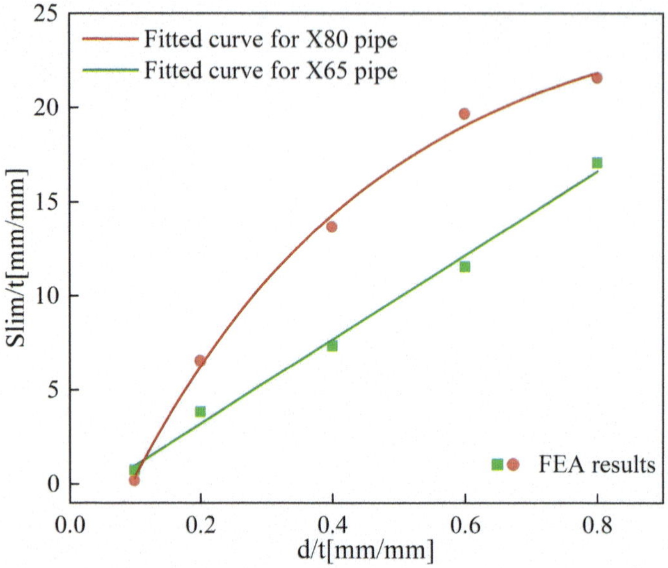

Fig. 2.8 Fitted curves for normalized limit spacing distance with different normalized corrosion depths

corrosion defects. It integrates a series of numerical approaches, including the FEM, MCS, sensitivity analysis, and artificial neural network method. The flowchart of the proposed reliability method estimating the reliability of the corroded pipeline considering the interacting effect is shown in Fig. 2.9. This framework has six steps, which can be divided into two major procedures: interaction rule development (Steps 1–2) and reliability analysis (Steps 3–6).

Once failure criteria are determined and the new interaction rule is developed, they can be used in burst pressure estimation and reliability analysis.

2.6.1 Data Sets Generation

Pipeline failure due to corrosion is a rare event with limited in-line inspection data for reliability forecasts, necessitating statistical methods to compile adequate data. Monte Carlo simulation (MCS), a proven reliable algorithm for reliability analysis, generates necessary datasets for burst pressure estimation through random sampling from parameter probability distributions, as depicted in Fig. 2.10. These datasets feed into a burst pressure model, with the output assessed against a limit state function to determine pipeline failure.

The database generated for the reliability analysis should be separated into two groups according to the limit spacing distance estimated by the interaction rule. If the generated spacing distance l_s is within the limit distance S_{lim}, the data set will

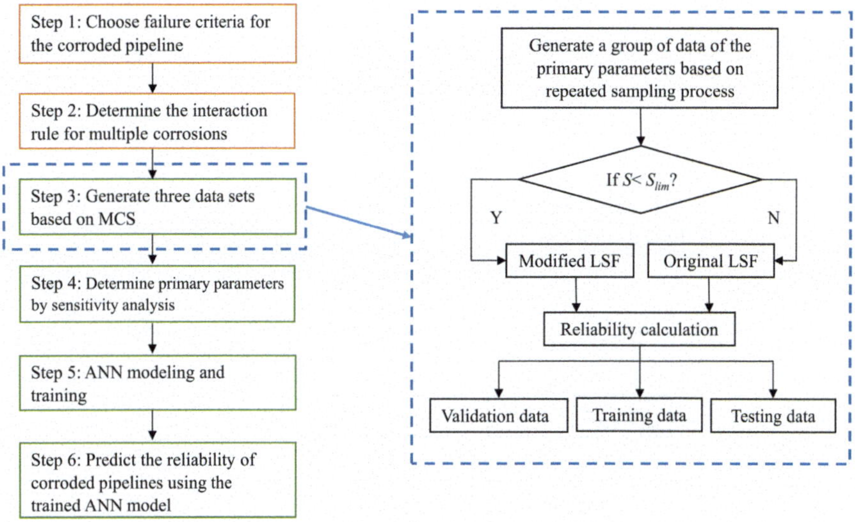

Fig. 2.9 Proposed method for reliability assessment of pipelines with multiple corrosion defects

be classified into the group in which the modified LSF is applied to consider the interacting effect. Otherwise, the data set will be classified into the group in which the original LSF is applied. In the modified LSF, the burst pressure is estimated by the burst model that incorporates the impairment of residual strength caused by the interacting effect. Spacing distance is usually defined as the distance between the edges of adjacent corrosion defects. The equivalent corrosion length and depth expressed by Eqs. 2.23 and 2.24, respectively. They are employed to substitute for the original ones in the burst model to quantify the interacting effect, as shown in Fig. 2.11. The corrosion depth used is the maximum depth of the corrosion defect [20].

$$l_e = l_1 + l_2 + l_s \tag{2.23}$$

$$d_e = \frac{l_1 d_1 + l_2 d_2}{l_e} \tag{2.24}$$

where l_e is the equivalent corrosion, l_1 and l_2 is the length of corrosion 1 and 2, respectively, d_1 and d_2 is the depth of corrosion 1 and 2, respectively.

In practice, pipe wall corrosion defects expand axially and radially due to environmental factors, which must be factored into reliability assessments. Corrosion growth is quantifiable by Eqs. 2.25 and 2.26 as per prior studies [20].

$$l(t) = l_0 + V_l(T_c - T_0) \tag{2.25}$$

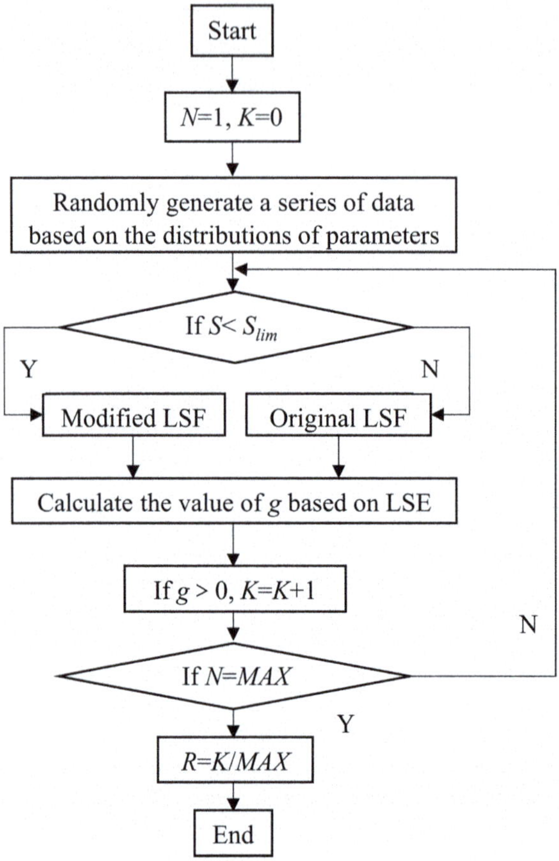

Fig. 2.10 Reliability prediction by MCS

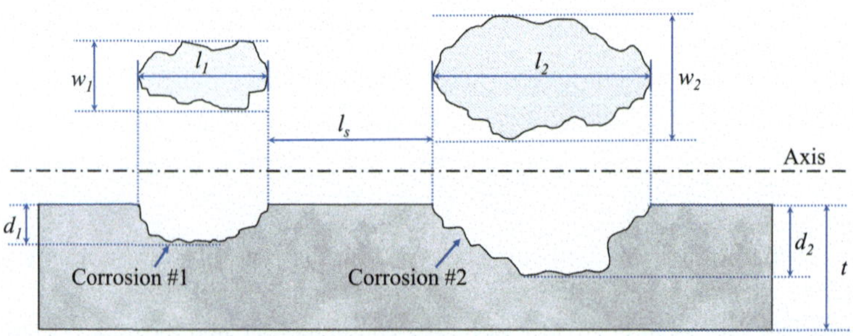

Fig. 2.11 A pipeline with interacting corrosion defects

$$d(t) = d_0 + V_d(T_c - T_0) \tag{2.26}$$

where $l(t)$ is the current corrosion length; l_0 is the initial corrosion length, V_l is the corrosion growth rate along the length, T_c is the current time, T_0 is the initial time, $d(t)$ is the current corrosion depth, d_0 is the initial corrosion depth, and V_d is the corrosion growth rate along the depth.

The database created via MCS is divided into training, validation, and testing sets for ANN, with the distribution percentages tailored to specific needs. ANN inputs are the sampled parameters, and its outputs are the reliabilities calculated by MCS.

2.6.2 Primary Parameters Determination

The complexity and computation time of an artificial neural network (ANN) are heavily influenced by its input parameters, typically derived from the burst pressure model's geometric and material parameters, as well as operational parameters. While more parameters can enhance ANN prediction accuracy, they also complicate the ANN structure and reduce computational efficiency. To balance efficiency and prevent overfitting, parameter quantity must be judiciously chosen without arbitrarily reducing input neurons, which could degrade ANN performance. The Sobol method, a variance-based approach, is employed in reliability analysis to evaluate the influence of each parameter on reliability predictions, handling nonlinearities effectively. Through the Sobol method, two Sobol indexes, S_i and S_{Ti}, can be obtained to evaluate the estimated reliability's sensitivity to the parameters. The functions of these two indexes are expressed by Eqs. 2.27 and 2.28, respectively. The total-effect index is generally used in sensitivity analysis to represent sensitivity.

$$S_i = \frac{\mathrm{var}_{Xi}\left(E_{X_{\sim i}}(Y|X_i)\right)}{\mathrm{var}(Y)} \tag{2.27}$$

$$S_{Ti} = \frac{E_{X_{\sim i}}(\mathrm{var}_{Xi}(Y|X_{\sim i}))}{\mathrm{var}(Y)} \tag{2.28}$$

where S_i is the first-order sensitivity index, X are input parameters, Y is the output, and S_{Ti} is the total-effect index.

2.6.3 ANN Modeling and Training

Traditional linear regression methods fall short in analyzing pipeline reliability with multiple corrosion defects due to complex, nonlinear parameter relationships. Artificial neural networks (ANN) excel in handling such complexities, effectively estimating pipeline reliability by mapping primary parameters as input neurons to output reliability. However, disparities in input magnitudes can desensitize the ANN to smaller-scale parameters. The corrosion depth is of the order of magnitude of 10, for example, while the pipe steel's ultimate tensile strength is of the order of magnitude of 10^8. Therefore, a feature scaling approach is required to ensure parameters or features are on a similar scale. The function of the feature scaling is expressed by Eq. 2.29. Through the feature scaling process, each parameter's range will be adjusted from − 1 to 1.

$$y = (y_{max} - y_{min}) \times \frac{(x - x_{min})}{(x_{max} - x_{min})} + y_{min} \quad (2.29)$$

where y is the value of x after feature scaling, x_{min} and x_{max} is the minimum and maximum value of x vector, respectively.

2.6.4 Conduct Reliability Prediction

The established and validated ANN serves as an efficient tool for rapidly estimating the reliability of pipelines affected by multiple corrosion defects.

2.7 Case Study

This section presents an example of an X65 pipeline with two axially aligned corrosion defects to demonstrate the proposed integrated reliability method. The proposed methodology aims to estimate the corroded pipeline's reliability considering the interacting effect over the operating lifetime. The probability distributions of the major parameters in an initial state are listed in Table 2.5. Each parameter will be used in the limit spacing distance estimation, burst pressure assessment, MCS, and ANN training. It is assumed that two corrosion defects are on the same pipe segment and in the same corrosive environment, i.e., the corrosion growth rates are the same for the defects. The spacing distance between the longitudinally aligned corrosion defects is 100 mm in the initial state, and it will be shortened because of the corrosion growth. Additionally, some geometric and material parameters remain stable during the operating period, such as the ultimate tensile strength and the outer diameter.

In Step 1, the LSF expressed by Eq. 2.19 is selected as the failure criterion of the X65 corroded pipeline. The calibrated burst pressure used in the function can be estimated by Eq. 2.17. Because X65 is medium-grade steel, X_{corr} is 1.21 [21].

In Step 2, FEM is employed to develop the interaction rule. Detailed information is presented in Sect. 2.5. The new interaction rule described in Eq. 2.21, which considers the effects of the corrosion depth and X65 pipe steel, is applied to determine the limit spacing distance between corrosion defects in the following step.

Table 2.5 Probability distributions of parameters

Parameters	Mean value	COV (%)	Distribution
UTS (MPa)	565	7	Normal
D (mm)	458.8	3	Normal
DTR	0.3	10	Normal
t (mm)	10	5	Normal
P (MPa)	8	8	Gumbel max
L (mm)	52	50	Normal
V_d (mm/year)	0.097	75	Normal
V_l (mm/year)	3	13	Normal

In Step 3, MCS is used to generate the reliability data sets. The process of MCS is shown in Fig. 2.10. The number of sampling times influences the accuracy and efficiency of MCS. Therefore, choosing an appropriate number of sampling times is significant. Based on the probability distributions listed in Table 2.5, the efficiency and stability analysis is carried out to investigate the effects of the number of sampling times on the MCS results and computing efficiency. Figure 2.12 shows comparisons of boxplots, skewness, and elapsed time needed for simulation under different numbers of sampling times. When the number of sampling times is less than 10^6, although the elapsed time is relatively short, the skewness, median value, and range within 1.5 times the interquartile range fluctuate with the variation of the number of sampling times. If the number of sampling times is greater than or equal to 10^6, the elapsed time of MCS shows a significant rise with the increase in the sampling times. However, the skewness and boxplot are stable. Therefore, 10^6 was set as the number of sampling times in this chapter. A data set was generated through MCS, including input data and target data for ANN training.

Based on MCS, Fig. 2.13 shows the reliability of the X65 pipeline with multiple corrosion defects under different interaction rules. Over 30 years, reliability declines at varying rates. Before 15 years, it remains stable regardless of interaction rules, but after that, CW and the new rule show the earliest decline, followed by single corrosion and ASME (~ 4 years later). Before 22 years, ASME treats defects as isolated. Underestimating interaction effects may delay maintenance, increasing accident risks. When reliability reaches 0.9 (maintenance threshold), the time lag between single and interacting corrosion is over 7.5 years, and between ASME and the new rule is about 1 year. The new rule, specifically developed for this case, ensures more accurate reliability predictions for corroded X65 pipelines.

Because of unstable flow conditions or significant changes in demand of the conveying media, pressure fluctuations might occur that could impact the corroded pipeline's reliability. The Poisson square wave process (PSWP) can be applied to simulate the internal pressure considering pressure fluctuations. Detailed information about the simulation process can be found in Tee and Pesinis [17]. Because of the inadequacy of time history for the internal pressure, the generation rate (λ) in the PSWP model is assumed as 0.75, 1.0, and 1.5. Figure 2.14a shows the simulated

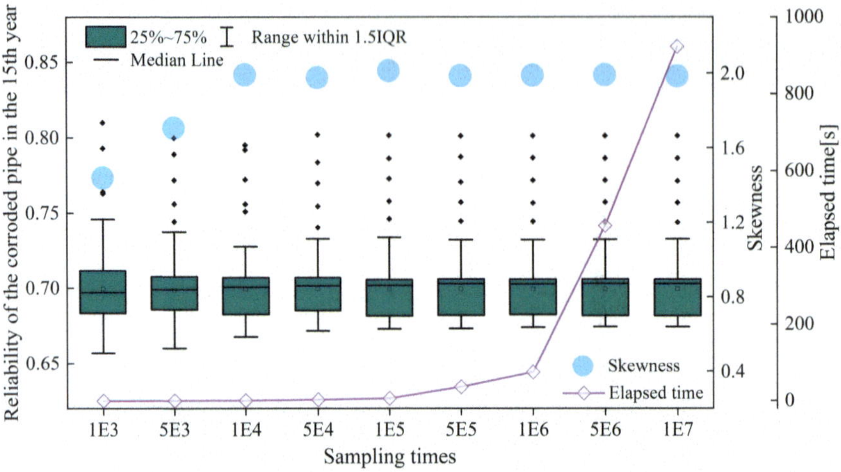

Fig. 2.12 Decision on the sampling times of the Monte Carlo simulation

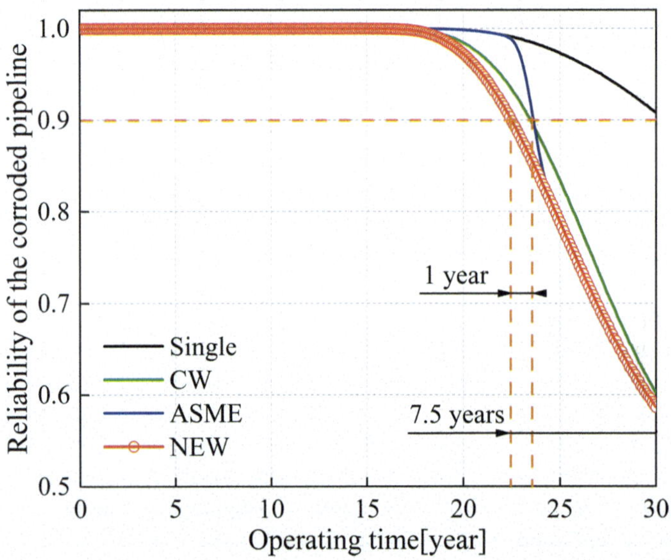

Fig. 2.13 Reliability predictions with different interaction rules versus operating time

pressures under different λ values. Figure 2.14b shows reliability predictions under different internal pressures, including PSWP and RV simulations. Pressure fluctuations have minimal impact near the maintenance threshold, so they are excluded from the proposed method. Future work may incorporate them once a real λ is obtained.

In Step 4, the Sobol method is implemented to determine primary inputs for further ANN construction. Based on previous simulation examples, ranges of parameters are

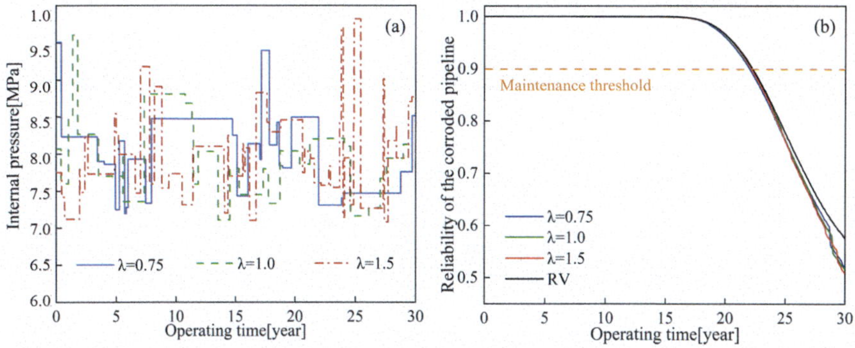

Fig. 2.14 Effects of pressure fluctuations on reliability of the corroded pipeline **a** simulated internal pressure based on PSWP **b** reliability predictions with different internal pressures

as follows: P = 6–14 MPa, t = 9.5–10.5 mm, DTR = 0.3–0.6, l = 50–150 mm, UTS = 525–605 MPa, D = 445–473 mm, V_d = 0.05–0.2 mm/year, and V_l = 2–6 mm/year. The time is from 0 to 30 years. Equation 2.15 total-effect index represents each parameter's contribution, with Sobol analysis results shown in Fig. 2.15. Operating time has the highest impact, while outer diameter has the least, with D and UTS contributing only 0.83%. Since at least 99% of the global contribution is retained for accuracy, outer diameter variation can be ignored in further reliability analysis. The selected primary parameters for the reliability analysis are the time, P, t, DTR, l, V_d, and V_l.

In Step 5, a feed-forward BP ANN is built and trained using a 9000-point dataset, split into training (70%), validation (15%), and testing (15%). The three-layer ANN structure (Fig. 2.16) uses input neurons from Step 4's sensitivity analysis and outputs

Fig. 2.15 Sobol analysis results of the parameters

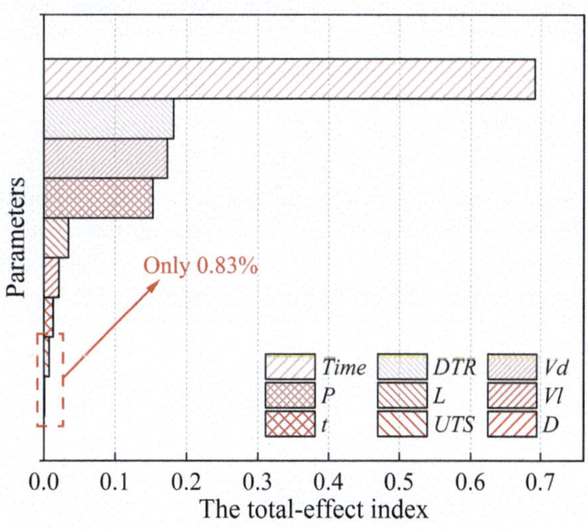

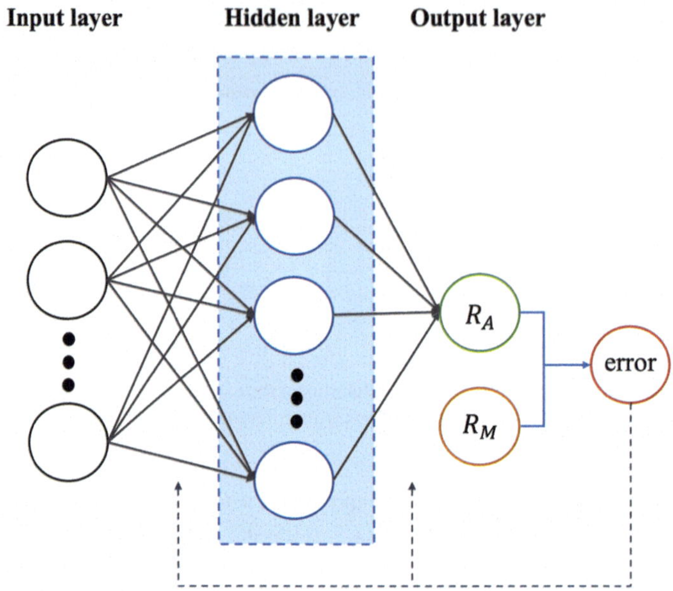

Fig. 2.16 ANN model architecture

pipeline reliability. Before training, inputs are scaled using Eq. 2.16.The best validation performance of the trained ANN is 8.37×10^{-6}. Moreover, the mean-square error (MSE) of the testing data set (15%) is around 8.51×10^{-6}, which is very small. Thus, the performance of the trained ANN is excellent.

The trained ANN predicts pipeline reliability with multiple corrosion defects. Figure 2.17 compares ANN and MCS predictions over 30 years, with 60 evenly spaced points per curve. ANN estimates closely match MCS results, with a mean-square error of 2.4×10^{-6}. MCS, a common validation reference, confirms ANN's accuracy. ANN requires only 1.08 s in MATLAB, compared to 52.28 s for MCS, demonstrating significantly higher efficiency.

Assessing pipeline reliability with multiple adjacent corrosion defects using the trained ANN offers significant advantages. It ensures satisfactory accuracy, high computational efficiency, and easier handling of nonlinear parameter relationships. The ANN's flexibility allows updates with new data, enhancing its reliability. This makes it a powerful tool for reliability analysis and maintenance management, especially for complex corrosion conditions.

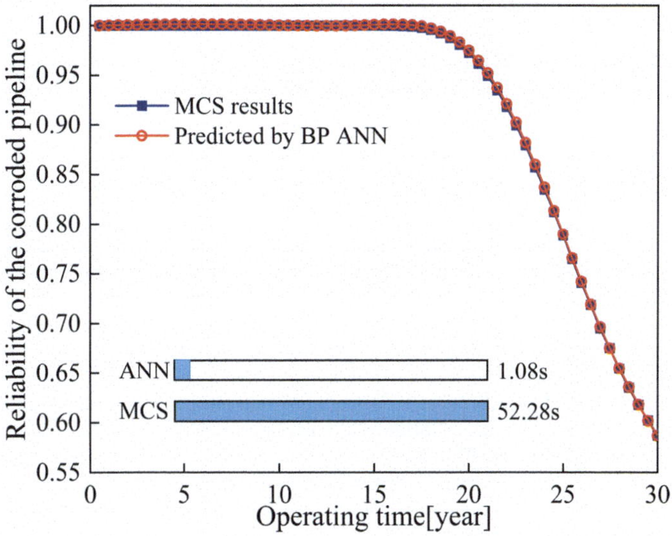

Fig. 2.17 Comparison of the reliabilities assessed by MCS and the trained ANN

2.8 Conclusions

This chapter presents an integrated approach for analyzing the reliability of pipelines with multiple corrosion defects, focusing on interaction effects. The approach involves two main steps: interaction development and reliability analysis. It was found that corrosion depth significantly influences interaction effects and burst pressure, with deeper corrosion intensifying defect interactions, especially in X80 steel pipelines. Traditional interaction rules were deemed inadequate as they overlooked corrosion depth and steel grade, leading to the development of two new rules for more accurate predictions.

The study highlights the importance of interaction effects in maintenance scheduling, with potential delays exceeding 7.5 years if not considered. A newly developed interaction rule provides more precise reliability assessments. An Artificial Neural Network (ANN) was trained, excluding the least impactful parameter, pipeline diameter, resulting in high prediction accuracy and computational efficiency, with a mean squared error of just 2.4×10^{-6} and a computation time of 1.08 s, far outpacing traditional methods.

The methodologies and outcomes of this research contribute to better integrity management of corroded pipelines, with suggestions for future studies to include environmental influences for a more comprehensive reliability analysis.

This chapter is adapted from the article 'An Integrated Reliability Method with a Newly Developed Interaction Rule for Steel Pipelines with Multiple Corrosion Defects,' originally published in *Journal of Pipeline Systems Engineering and Practice*, 2022, DOI: https://doi.org/10.1061/(ASCE)PS.1949-1204.0000683. Reproduced with permission from ASCE.

References

1. Zhang J, Liang Z, Han J (2014) Failure analysis and finite elementsimulation of above ground oil–gas pipeline impacted by rockfall. J Fail Anal Prev 14(4):530–536. https://doi.org/10.1007/s11668-014-9847-x
2. Wasim M, Djukic MB (2022) External corrosion of oil and gas pipelines: a review of failure mechanisms and predictive preventions. J Nat Gas Sci Eng 100:104467. https://doi.org/10.1016/j.jngse.2022.104467
3. Wu W et al (2018) Electrochemical and stress corrosion mechanism of submarine pipeline in simulated seawater in presence of different alternating current densities. Materials 11(7):1074. https://doi.org/10.3390/ma11071074
4. Engineers AS of M (2012) Manual for determining the remaining strength of corroded pipelines: a supplement to ASME B31 code for pressure piping: an American national standard. American Society of Mechanical Engineers
5. Kiefner JF, Vieth PH (1989) A modified criterion for evaluating the remaining strength of corroded pipe. Battelle Columbus Div., OH (USA)
6. Association CS et al (2015) Oil and gas pipeline systems-CSA Z662-15. CSA Group
7. Det Norske Veritas AS (2019) Corroded pipelines: recommended practice DNVGL-RP-F101. Edition Sept 2019
8. Leis BN, Stephens DR (1997) An alternative approach to assess the integrity of corroded line pipe. Part 1: current status
9. Abdelmoety AK, Zheng Q, Li Y, Kainat M, Yoosef-Ghodsi N, Adeeb S (2021) Probability of failure associated with design and safety factors for intact and corroded pipes under internal pressure. J Pipeline Syst Eng Pract 12:04021010. https://doi.org/10.1061/(ASCE)PS.1949-1204.0000544
10. Nanninga NE et al (2012) Comparison of hydrogen embrittlement in three pipeline steels in high pressure gaseous hydrogen environments. Corros Sci 59:1–9. https://doi.org/10.1016/j.corsci.2012.01.028
11. Chen K, Ma H, Che L, Li Z, Wen B (2019) Comparison of meshing characteristics of helical gears with spalling fault using analytical and finite-element methods. Mech Syst Signal Process 121:279–98. https://doi.org/10.1016/j.ymssp.2018.11.023
12. Kiefner JF, Vieth PH (1990) PC program speeds new criterion for evaluating corroded pipe. Oil Gas J USA 88:34. https://www.osti.gov/biblio/6828231
13. Chebaro MR, Zhou W (2010) A limit state function for pipelines containing long corrosion defects, pp 511–517. https://doi.org/10.1115/IPC2010-31324.
14. Bhardwaj U, Teixeira AP, Guedes Soares C (2020) Uncertainty quantification of burst pressure models of corroded pipelines. Int J Press Vessels Pip 188:104208. https://doi.org/10.1016/j.ijpvp.2020.104208
15. Sun J, Cheng YF, Woo J, Kainat M, Hassanien S (2021) Assessment of interaction between a dent and an adjacent corrosion feature on pipelines and the effect on pipeline failure pressure by finite-element modeling. J Pipeline Syst Eng Pract 12:04021029. https://doi.org/10.1061/(ASCE)PS.1949-1204.0000570
16. Zhang S, Zhou W (2021) Development of a burst capacity model for corroded pipelines considering corrosion defect width and a revised Folias factor equation. J Nat Gas Sci Eng 88:103812. https://doi.org/10.1016/j.jngse.2021.103812
17. Bao J, Zhou W (2021) Influence of depth thresholds and interaction rules on the burst capacity evaluation of naturally corroded pipelines. J Pipeline Sci Eng 1:148–65. https://doi.org/10.1016/j.jpse.2021.01.001

18. Netto TA, Ferraz US, Estefen SF (2005) The effect of corrosion defects on the burst pressure of pipelines. J Constr Steel Res 61:1185–1204. https://doi.org/10.1016/j.jcsr.2005.02.010
19. Sun J, Cheng YF (2018) Assessment by finite element modeling of the interaction of multiple corrosion defects and the effect on failure pressure of corroded pipelines. Eng Struct 165:278–86. https://doi.org/10.1016/j.engstruct.2018.03.040
20. Zhang S, Zhou W (2022) Assessment of the interaction of corrosion defects on steel pipelines under combined internal pressure and longitudinal compression using finite element analysis. Thin-Walled Struct 171:108771. https://doi.org/10.1016/j.tws.2021.108771
21. Mondal BC, Dhar AS (2019) Burst pressure assessment of corroded pipelines using fracture mechanics criterion. Eng Fail Anal 104:139–153. https://doi.org/10.1016/j.engfailanal.2019.05.033

Chapter 3
Damage Mechanism Investigation of Offshore Pipelines Impacted by Falling Objects

3.1 Introduction

This chapter investigates the vulnerability of offshore pipelines to impacts from falling objects, a significant threat given their typical exposure on the seabed. Two key studies are presented: the first assesses the damage from bar-shaped objects, using numerical simulations to understand the impact energy distribution and mechanical failure risk, particularly on rock seabed. The second study examines lump-shaped objects' impacts on pipelines, exploring how seabed soil how seabed soil mitigates impact forces and how various factors influence the extent of the damage. These insights are crucial for enhancing the design and safety protocols of offshore pipelines to prevent environmental and economic disasters.

3.2 Theoretical Method

3.2.1 *Energy Method*

The oceanic object descent is complex, but for practicality, DNV-RP-F107 [1] breaks it down into two stages. Initially, objects accelerate as water resistance is less than buoyancy (Stage 1). Eventually, a constant velocity is achieved when forces balance out, typically 50–100 m deep (Stage 2). For this analysis, impact velocity is considered as the maximum constant velocity from Stage 2, as per the force-balance principle.

$$mg - \rho_w Vg - \frac{1}{2} C_d \rho_w A v_m = 0 \tag{3.1}$$

H. Zhang et al., *Submarine Pipeline Integrity: Assessment of Failure Modes and Advanced Evaluation Techniques*, Ocean Engineering & Oceanography 28,
https://doi.org/10.1007/978-3-031-92092-9_3

$$v_m = \sqrt{2mg\frac{\left(1 - \frac{\rho_w}{\rho_i}\right)}{C_d \rho_w A}} \quad (3.2)$$

where m is the mass of the falling object (kg); g is the gravitational acceleration (m/s^2); ρ_w is the mass density of the seawater (kg/m^3); V is the volume of the falling object (m^3); ρ_i is the mass density of the falling object; C_d is the drag coefficient; A is the projected area of the object in the flow direction (m^2).

When a large body moves through a fluid, it drags some of the fluid along due to inertia, creating additional resistance. To account for this, the concept of 'added mass' is used, which quantifies the effect of the fluid's inertia based on the body's motion and shape. For the bar-shaped objects studied here, the interaction with seawater is significant and must be factored in. The added hydrodynamic mass is typically incorporated as an equivalent mass density for the falling object.

$$\rho_e = \rho_i\left(1 + \frac{\rho_w}{\rho_i}C_a\right) \quad (3.3)$$

where ρ_e is the equivalent mass density of the falling object (kg/m^3); C_a is the added mass coefficient.

The drag coefficient can reflect the interaction between seawater and the falling object C_d and added mass coefficient C_a, which both depend on the geometrical features of the falling object. The relation between the shape of the falling object and corresponding values of C_d and C_a are given in Table 3.1.

Due to the conservation of the energy in the closed system, the impact energy of the falling object is the total kinetic energy, which is accumulated in stage 1 and maintained in stage 2. The following formula can express the impact energy:

$$E_i = \frac{1}{2}m_e v_m^2 \quad (3.4)$$

It is noted in DNV-RP-F107 that falling objects can be assumed to have an impact velocity defined by Eq. 3.2 when the depth is below 50 m. Based on Eqs. 3.2 and 3.4, the impact energy of falling objects can be deduced as Eq. 3.5. It is obvious that this energy equation is suitable for the water depth larger than 50 m.

Table 3.1 Typical values of the drag coefficient C_d and added mass coefficient C_a

The shape of the falling object	Added mass coefficient C_a	Drag coefficient C_d
Box shape	$1.2 \le C_a \le 1.3$	$0.6 \le C_d \le 1.5$
Slender shape	$0.7 \le C_a \le 1.5$	$0.1 \le C_d \le 1.0$
Spherical shape to complex	$0.6 \le C_a \le 2.0$	$1.0 \le C_d \le 2.0$

$$E_i = \frac{(\rho_i^2 - \rho_w^2)m^2 g C_a}{\rho_i^2 \rho_w A C_d} \tag{3.5}$$

where E_i is the impact energy of the falling object; m_e is the equivalent mass of the falling object includes added hydrodynamic mass (kg/m^3).

3.2.2 *Pipeline Impact Calculation*

(1) ***Bai Yong empirical formula*** [2]

Based on the linear elastic solution, numerical, and experimental results provided by previous research, Bai developed an equation, which is expressed as Eq. 3.6, to describe the nonlinear relation between the load and linear elastic displacement of the load point on the circular thin-walled tubes. This equation is suitable for calculating the local indentation on tubular structure.

$$\delta_E = 0.116\frac{F}{L_c E}\left(\frac{D}{t}\right)^3 \tag{3.6}$$

where D is the external diameter of the pipeline (m); F is the impact load (N); E is Young's modulus; t is the wall thickness of the pipeline (m); δ_E is the elastic deformation of the pipeline at the load point; L_c is the characteristic length of the contact area along the axial direction of the pipe.

According to linear finite shell element analysis and indentation tests, an empirical equation estimating L_c can be expressed as Eq. 3.7.

$$L_c = 1.9D \tag{3.7}$$

The critical load is a measurement used to determine whether a permanent indentation will occur on the tubes. Through a rigid-plastic analysis, the equation for describing the critical load is derived, as expressed by Eq. 3.8.

$$F_c = 2\sigma_y t^2 \frac{L_c}{D} \tag{3.8}$$

where F_c is the critical load (N); σ_y is the yield stress of the material.

(2) ***A force-dent proposal proposed by Mario A. Polanco-Loria*** [3]

Based on several classical empirical equations and the observations from the finite element analysis, Mario A. Polanco-Loria introduced a new analytical expression to describe the relationship between the impact force and the dent of pipelines. The relationship between the force and the plastic dent can be expressed as Eq. 3.9.

$$h_p = \frac{0.25F^3}{K^s\left(F_Y^s\right)^2} \tag{3.9}$$

where h_p is the permanent dent depth, F is the indentation force, K^s is the steel pipe stiffness, F_Y^s is the denting force that causes yield.

According to DNV-RP-F111 and the simulation results, the flow stress concept is introduced to modify the equation of the denting force. The modified denting force can be expressed as Eq. 3.10.

$$F_Y^s = 0.35\overline{\sigma}t^2(D/t) \tag{3.10}$$

where $\overline{\sigma}$ is the flow stress.

The stiffness of the steel pipeline can be obtained, expressed as Eq. 3.11, based on the numerical simulation of the unloading response of the pipe.

$$K^s = \frac{2600\sigma_y t}{(D/t)^{1.7}} \tag{3.11}$$

where σ_y is the yield stress of the pipeline steel.

3.3 Numerical Simulation Model

Offshore pipelines will be forced to dent and even totally damaged by the impact of falling objects caused by third-party activities. Low-velocity impact problems involve several complex nonlinearities, such as geometrical nonlinearity, material nonlinearity, etc. Theoretical models are not suitable for such challenging situations [4]. The finite element method (FEM) has been widely used to solve complex engineering problems over the past decades. According to previous research, FEM could obtain accurate simulation results with relatively high computing efficiency when solving impact problems [5]. Thus, FEM (dynamic, explicit) is employed in this paper to investigate the mechanical behaviors of the offshore pipeline impacted by the falling object. As shown in Fig. 3.1, the 3D finite element model consists of a bar-shaped or lump-shaped falling object, an offshore pipeline, and a seabed.

To improve computational efficiency and reduce costs, a half-model is used due to the symmetry of both the structure and impact load. A 15 m pipeline length (L = 15 m) is selected to minimize end effects and enhance simulation accuracy. Based on offshore pipeline design standards, the single-layer pipeline has a diameter of 0.508 m (D = 0.508 m) and a wall thickness of 0.0125 m (t = 0.0125 m). The seabed absorbs part of the impact energy, with its cubic dimensions set at H = 2 m, L = 15 m, and W = 3 m. The soil foundation bottom is fixed, with symmetry applied along the Z-direction (Fig. 3.1). The bar-shaped falling object measures 2 m in length, 0.4 m in width, and 0.2 m in height. Additionally, a cylindrical lump-shaped object with a

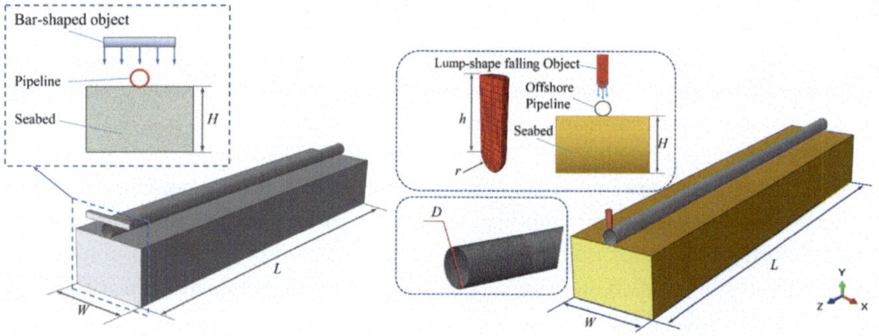

Fig. 3.1 Finite element model of offshore pipeline impacted by the bar-shaped and lump-shaped falling object

spherical impact end has a 0.5 m cylindrical height ($h = 0.5$ m) and a 0.1 m spherical radius ($r = 0.1$ m). The pipeline is meshed with 4-node shell elements, while the soil uses 8-node hexahedron elements. Mesh refinement is applied near the impact zone for higher accuracy.

To enhance computational efficiency and reduce costs, a half-model (1/2 model) is utilized for calculations due to the symmetry of both the geometric structure and the impact load from the falling object. The pipeline length is crucial for the economic feasibility of the simulation. A length of 15 m ($L = 15$ m) is chosen to negate end effects and increase the precision of the simulation results. According to the existing offshore pipeline design manual, the geometric parameters of the single-layer offshore pipeline are selected as follows: the pipeline's diameter is 0.508 m ($D = 0.508$ m), and the wall thickness is 0.0125 m (t = 0.0125 m). During the impact process, the seabed could buffer and absorb part of the impact energy. The geometric parameters of the cubic seabed are: the height equals 2 m (H = 2 m), the length is 15 m ($L = 15$ m), and the width equals 3 m ($W = 3$ m). The bottom of the soil foundation is set fixed, and the symmetric boundary condition is set along the Z-direction, as shown in Fig. 3.1. The length of the bar-shaped falling object is set as 2 m, the width is 0.4 m, and the height is 0.2 m. The falling object is a cylindrical lump-shaped object which has a spherical impact end. The height of the cylindrical part is 0.5 m ($h = 0.5$ m), and the radius of the spherical part is 0.1 m ($r = 0.1$ m). 4-nodes shell elements are employed to mesh the submarine pipeline, while the 8-nodes and hexahedron elements are used to mesh the soil. To improve the calculating exactness, mesh refinement is needed for the areas near the impact zone [6].

The structural integrity and impact response of submarine pipelines is significantly influenced by the internal operating pressure and external water pressure. Gathering pipelines typically operate within a pressure range of 6.9–10 MPa, while some long-distance pipelines function at lower pressures, such as the CPOE's pipeline between artificial islands NP1-3D and NP1-1D, which operates at a maximum of 1.2 MPa. External pressure is determined by the depth of the water.

The submarine pipeline material is X65 steel, commonly used in oil and gas pipelines, with a mass density of 7850 kg/m^3, Young's modulus of 210 GPa, and a Poisson's ratio of 0.3 [7]. The plastic properties of X65 steel, including strain rate and strain hardening, significantly influence impact behavior.

For geotechnical modeling, the Mohr–Coulomb and Drucker-Prager/Cap models are widely used for pipeline-soil interactions. The Drucker-Prager/Cap model is selected due to its numerical stability and ability to capture the seabed's elastic–plastic deformation under impact [8]. Given the variability of seabed conditions, sandy soil is chosen, with a mass density of 1923 kg/m^3, Young's modulus of 5 MPa, and a Poisson's ratio of 0.3.

The added hydrodynamic mass of the falling object is considered, yielding an equivalent mass density of 8875 kg/m^3. The submarine pipeline is assumed to be 100 m below the sea surface, with the bar-shaped object unaffected by sea flow during impact.

Traditional empirical formulas for dent deformation estimation are unsuitable as they assume unpressurized, suspended pipelines. Instead, a numerical model based on Li's buried pipeline impact experiment is used to verify feasibility and reliability. The simulation results closely match experimental data, with strain discrepancies of 3.09% at the critical state and 3.46% at the final stage, confirming the model's accuracy in impact analysis.

3.4 Results and Discussion

3.4.1 Energy Analysis

During the impact of an exposed submarine pipeline by a bar-shaped object, energy dynamics among the pipeline, object, and seabed play a crucial role. With internal and external pressures of 1 MPa and an impact velocity of 5.482 m/s, energy changes are shown in Fig. 3.2a. Initially, the object's kinetic energy dominates, while the pipeline and seabed's internal energy is minimal. As impact progresses, the pipeline's internal energy rises faster than the seabed's until 0.02 s, when the object's kinetic energy depletes. Post-impact, the object may rebound, regaining some energy, while the seabed absorbs energy for about 0.04 s before stabilizing. The seabed ultimately absorbs the most energy, emphasizing the importance of pipeline-seabed coupling in impact studies.

Figure 3.2b illustrates energy distribution throughout the impact. Initially, the seabed and pipeline hold no energy until impact occurs. The object's kinetic energy, constituting the system's total pre-impact energy, drops to zero, transferring energy to the pipeline, which peaks rapidly. At the critical state, the object rebounds, recovering only 4.326% of its initial kinetic energy. The seabed absorbs 13.672% of the initial energy, stabilizing at 0.02 s. While the pipeline absorbs most energy, the seabed's contribution is essential for accurate impact simulations.

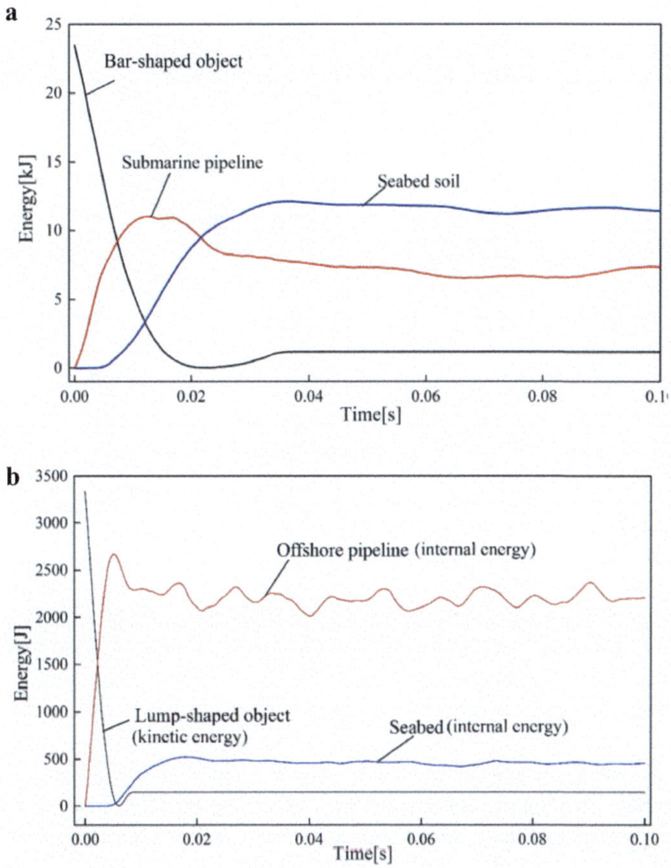

Fig. 3.2 Energy variation curves of each part of the numerical model along with time **a** bar-shaped object **b** lump-shaped object

3.4.2 Sensitivity Analysis for Bar-Shaped Object Impact

3.4.2.1 Type of Seabed

Submarine pipelines are laid on various seabed types, influencing their mechanical response to impact. To examine this effect, three seabed types—sand, soft soil, and rock—are modeled.

At an impact velocity of 5.382 m/s and internal/external pressures of 1 MPa, Fig. 3.3 presents the equivalent plastic strain and contour plots. Pipelines on sandy and soft soil seabeds show similar plastic strain, with deformation concentrated on the top. In contrast, pipelines on rock experience significantly higher strain, with an expanded plastic deformation area forming an 'm'-shaped pattern.

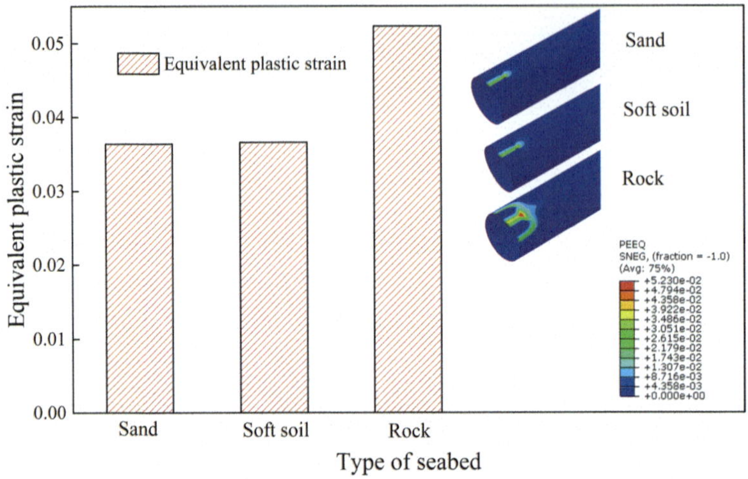

Fig. 3.3 Shows the absorbed energy proportions and the maximum depression rates of the submarine pipeline laid on different types of seabed

When the submarine pipeline is laid on the rock, most of the kinetic energy of the impact bar-shaped object is absorbed by the deformed pipeline, and the absorbed energy proportion of the pipeline exceeds 60%, which is the maximum among these three conditions. Unlike the other two types of seabed, rock is hard, and it is difficult for the rock to absorb the impact energy propagating from the pipeline through deformation. Due to the impact caused by the bar-shaped object, a dent could occur on the pipeline. The maximum depression rate of the pipeline laid on the rock is twice larger than that of the pipeline laid on the soft soil or sand (Fig. 3.4).

Fig. 3.4 Absorbed energy proportion and the maximum depression rate of submarine pipelines laid on different types of seabed

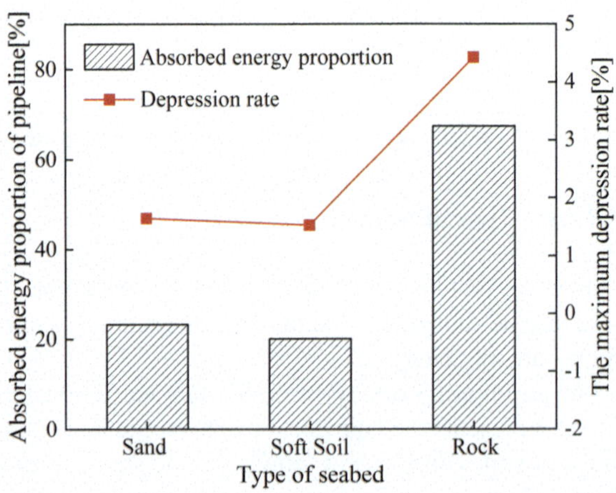

3.4.2.2 Impact Velocity

In most cases, bar-shaped objects falling from the sea surface have an initial velocity, such as those from offshore platforms. To analyze the effect of impact velocity, the object's volume, geometry, and density remain constant.

Figure 3.5 illustrates Von Mises stress distributions for different impact velocities. As velocity increases, the stress concentration zone expands axially and circumferentially. At velocities below 8 m/s, high stress appears on the pipeline top without significant denting. Above 8 m/s, stress concentrates at the dent's bottom, extending to the pipeline's sides. Higher velocities lead to more severe stress concentration.

As shown in Fig. 3.6, when the impact velocity of the bar-shaped object is lower than 6 m/s, plastic strain mainly located on the top of the pipeline, and the shape of the plastic strain area looks like a straight line. When the impact velocity is higher than 8 m/s, severe plastic strain takes place on the submarine pipeline's top, which distributes as "m" shape on the 1/2 pipeline model. Furthermore, not only appears in the center of the dent, but plastic deformation also occurs at the edge of the dent. The most severe plastic strain occurs on the arc ends of the m-shape plastic strain area when the impact velocity equals 14 m/s. Besides, both the equivalent plastic strain and the plastic strain area of the impacted submarine pipeline increase as the impact velocity increases.

Figure 3.7 presents pipeline depression rates and cross-sectional deformations for impact velocities from 4 to 14 m/s. Below 6 m/s, deformation is minimal with no visible dent, but structural reliability remains a concern due to environmental complexities. Sudden pressure differences could still lead to pipeline failure. Above 8 m/s, both depression rate and deformation increase with impact velocity. In the directly impacted area (0.0–0.2 m), depression rate remains stable, while beyond this zone (0.2–1 m), it gradually decreases with axial distance.

Figure 3.8 shows the absorbed energy proportion and maximum impact force of the pipeline at different impact velocities. Both exhibit a nonlinear increase with

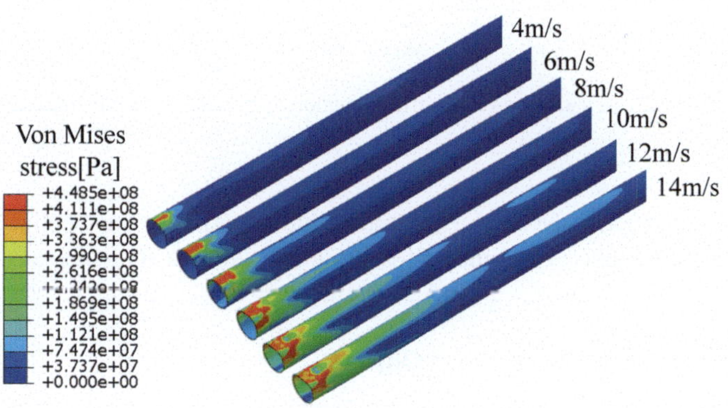

Fig. 3.5 Von Mises stress distributions in submarine pipelines under different impact velocities

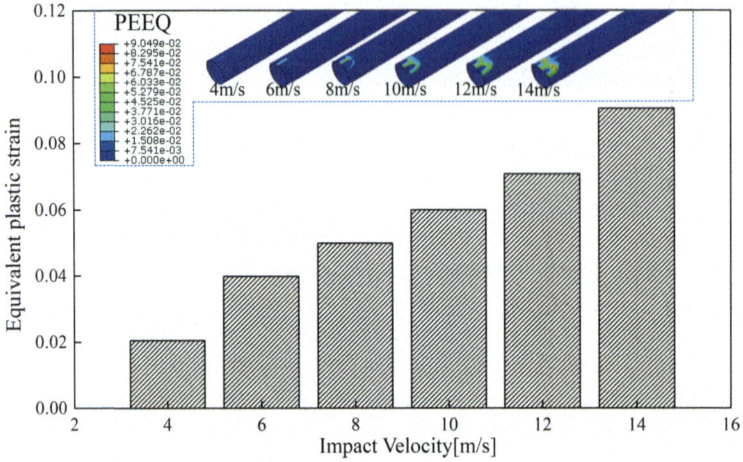

Fig. 3.6 Equivalent plastic strain and distribution in submarine pipelines under different impact velocities

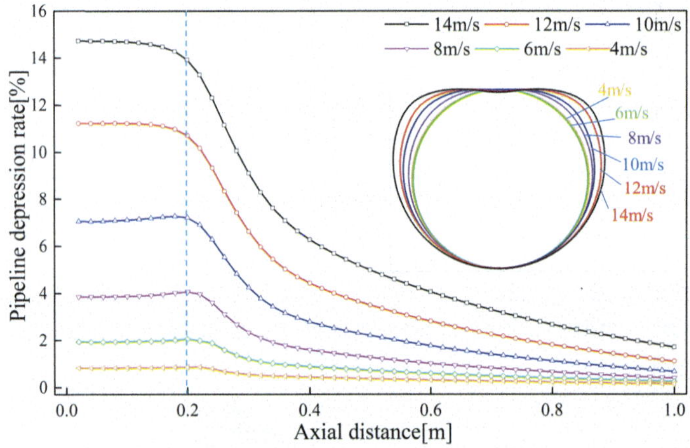

Fig. 3.7 Depression rate and cross-sectional deformation of the submarine pipeline under different impact velocities

velocity. Above 10 m/s, the energy absorption rate slows due to the pipeline's limited capacity, while the maximum impact force continues to rise at an increasing rate.

3.4.2.3 Shape of the Bar-Shaped Object

Submarine pipelines face impact risks from various falling objects, with different shapes causing varying mechanical damage. Studying these effects is crucial for

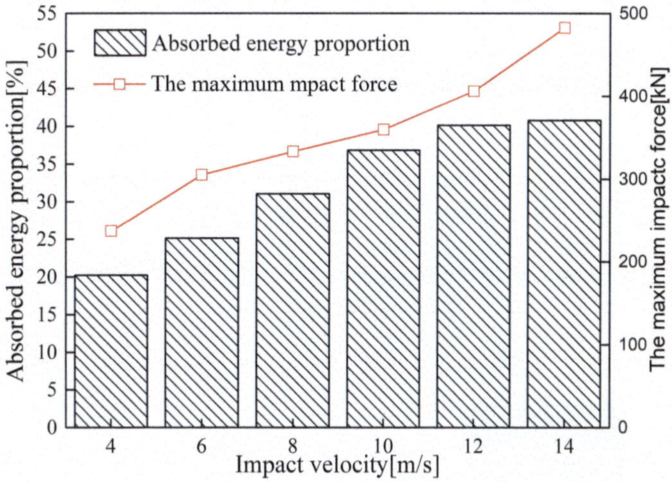

Fig. 3.8 Absorbed energy proportion and maximum impact force on submarine pipelines under different impact velocities

pipeline integrity management and protection system design. The shape of a falling object depends on ocean activity and is often simplified for analysis. This study examines three typical shapes—cubicle, cylindrical, and Tri-Prism (Fig. 3.9). All objects have identical projected areas in the flow direction, equal lengths and widths, and the same weight. Per Eq. 3.2, their impact velocity is 5.482 m/s.

At an internal and external pressure of 1 MPa and an impact velocity of 5.482 m/s, Fig. 3.10 shows Von Mises stress distributions for different bar-shaped object impacts. The Tri-Prism type, with a small contact area, causes stress concentration at the impact point and at the dent's edges. The cylindrical object produces a similar pattern but with a wider high-stress area. The cubic object mainly concentrates stress

Fig. 3.9 Bar-shaped objects with different shapes

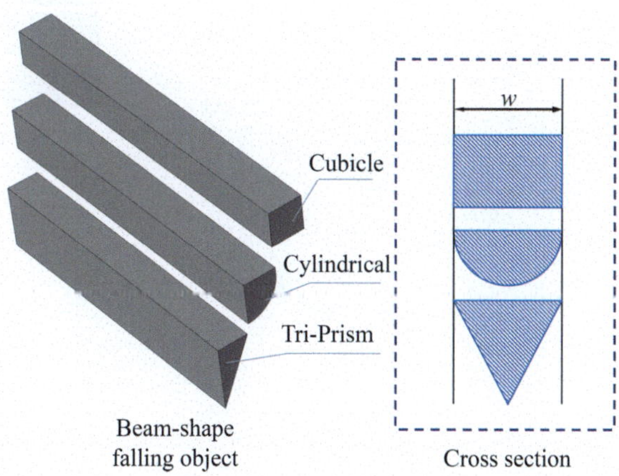

on the pipeline top, with less severity on the sides. Due to its larger contact area, the cubic object causes the least damage among the three types.

Figure 3.11 presents equivalent plastic strain distributions for impacts by cubicle, cylindrical, and Tri-Prism objects. The Tri-Prism type causes maximum plastic strain at the pipeline's edge, while the cylindrical object creates a slightly wider strain area along the axial direction. Both produce a hump-shaped strain pattern. The cubicle object results in the lowest strain, with plastic deformation confined to the pipeline top in all cases.

Figure 3.12 shows dent deformation along the axial direction. Depression decreases with axial distance. The cubicle object causes a lower depression rate than

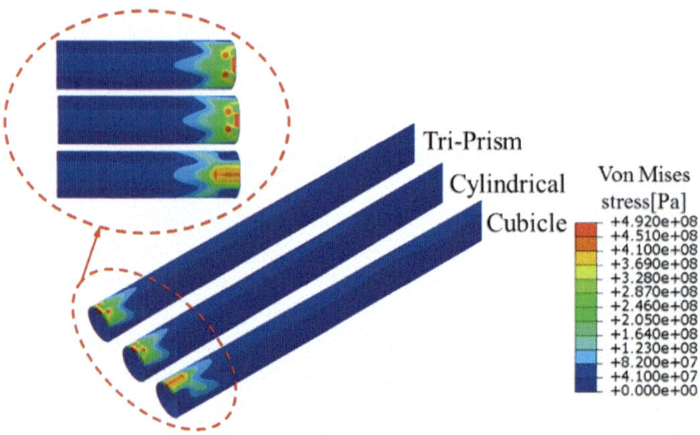

Fig. 3.10 Von Mises stress distributions of the submarine pipeline under different shapes of the bar-shaped object

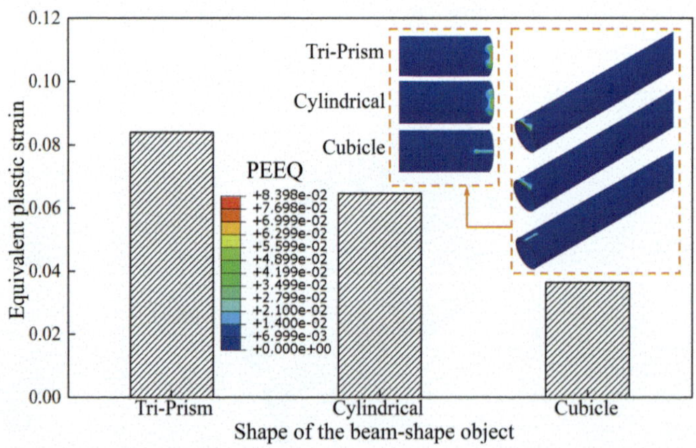

Fig. 3.11 Shows the curves of depression rates and cross-sectional deformations of the exposed submarine pipeline impacted by bar-shaped objects with different shapes

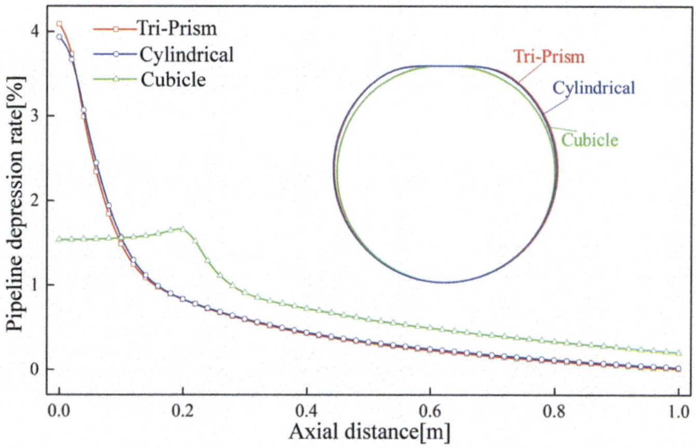

Fig. 3.12 Depression rate and cross-sectional deformation of the submarine pipeline under different shapes of the bar-shaped object

the Tri-Prism and cylindrical types within 0–0.1 m. For the latter two, the maximum depression occurs at the impact center (0 m), while for the cubicle type, it peaks at 0.2 m. Beyond the impact zone, the cubicle object results in a higher depression rate than the other two.

X65 submarine pipelines can convert the kinetic energy of the falling object into its internal energy. The proportions of the absorbed energy at the final state and the maximum impact forces of the submarine pipeline impacted by different shapes of bar-shaped objects are shown in Fig. 3.13. When the cubicle-type object impacts the submarine pipeline, the absorbed energy proportion (23.24%) of the pipeline is the minimum among these three different impact conditions. The maximum absorbed energy proportion of the submarine pipeline occurs when the Tri-Prism type object impacts the pipeline. Compared to the absorbed energy proportion, the maximum impact force of the submarine pipeline shows an opposite change rule. When the cubicle-type object impacts the pipeline, the maximum impact force is the highest, reaching 294.698kN.

3.4.2.4 Impact Angle

When a bar-shaped object falls in the ocean, its axial direction is not fixed and might change due to various external factors. Besides, submarine pipelines, which are not covered with buried soil or fixed by specific devices, are prone to change their designated positions because of environmental factors, such as ocean current, seabed landslide, etc. Thus, to make the research closer to the real situation, the impact angle between the bar-shaped object and the submarine pipeline is selected to be a significant research factor. The schematic diagram of the impact angle of the bar-shaped object is shown in Fig. 3.14. The impact angle α is defined as the angle

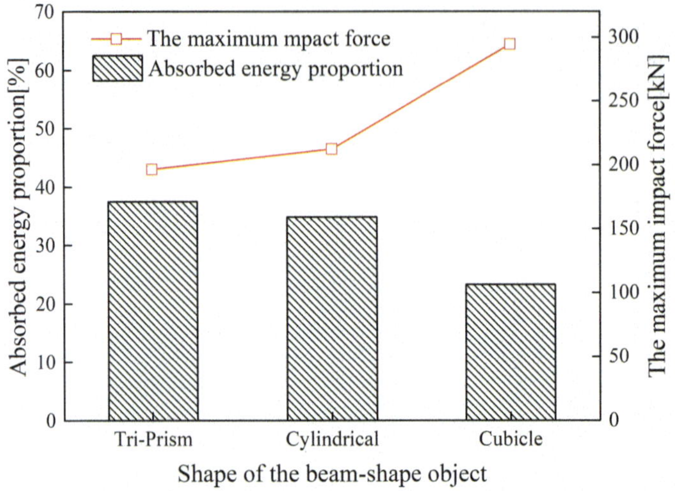

Fig. 3.13 Absorbed energy proportion and the maximum impact force of the submarine pipeline under different shapes of the bar-shaped object

between the axis of the bar-shaped object and the axis of the submarine pipeline; the range of the impact angle is from 0° to 90°.

Von Mises stress distributions of the submarine pipeline impacted by the bar-shaped object with different impact angles are shown in Fig. 3.15. With the increase of the impact angle, the high-stress area on the top of the pipeline gradually narrows along the axial direction; however, the stress concentration is getting progressively more and more serious. When the impact angle is between 30 and 45°, the stress concentration not only occurs at the top of the pipeline but also at the bottom of the pipeline. However, the stress concentration on the bottom is not as severe as that on the top. No matter how the impact angle changes, the stress concentration always

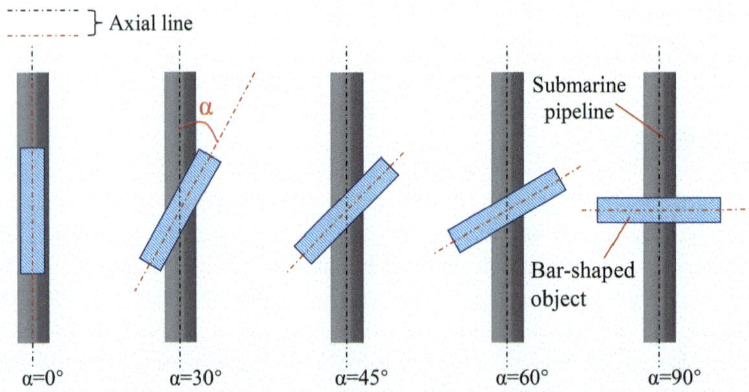

Fig. 3.14 Impact angle of the bar-shaped object

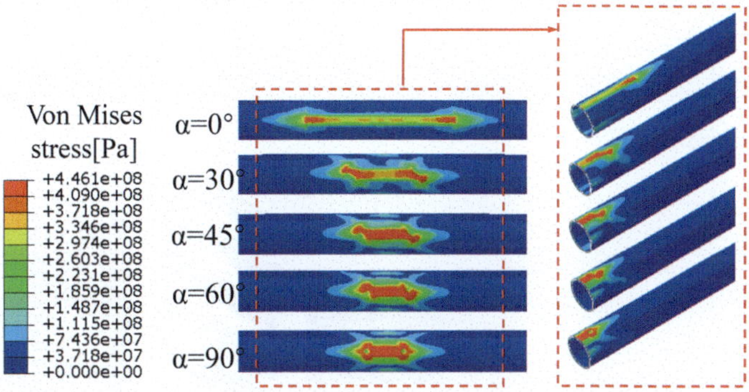

Fig. 3.15 Shows the equivalent plastic strain and distributions of the submarine pipeline under different impact angles

appears on the top of the submarine pipeline, and the most severe stress concentration occurs at the edge of the dent all the time.

Figure 3.16 shows equivalent plastic strain and distributions of the submarine pipeline under different impact angles. When the impact angle $\alpha \leq 45°$, the equivalent plastic strain increases with the increase of the impact angle. When the impact angle $\alpha \geq 45°$, the equivalent plastic strain of the submarine pipeline shows a downward trend with the rise of the impact angle. When the impact angle varies from 0° to 90°, the maximum equivalent plastic strain appears when the impact angle equals 45°. When $\alpha = 0°$, since the impact area is the largest, the plastic deformation of the submarine pipeline is not that severe. The plastic strain distributes as a straight line on the top of the pipeline. With the increase of the impact angle, the plastically deformed area gradually varies from axial distribution to circumferential distribution.

The impact angle of the bar-shaped object plays an essential role in the depression rate of the exposed submarine pipeline. Curves of the pipeline depression rates along with axial distance are shown in Fig. 3.17. When the impact angle equals 0°, the pipeline depression rates in the whole impact area (− 1 to 1 m) are almost all the same. The axial length of the dent caused by the bar-shaped object decreases with the growth of the impact angle. It can be inferred from the curves of the depression rates that the axial length of the dent decreases with the increase of the dent depth while the impact energies of the bar-shaped objects with different impact angles are all the same.

The maximum impact forces and the absorbed energy proportions of the submarine pipeline impacted by the bar-shaped object with different impact angles are shown in Fig. 3.18. The lowest absorbed energy proportion of the pipeline appears when the impact angle is equal to 0°; but, at the same time, the maximum impact force of this situation is the largest among the different impact angles. When the impact angle equals 90°, the maximum impact force of the submarine pipeline is the lowest, while the absorbed energy proportion is the highest, which reaches 16.97%.

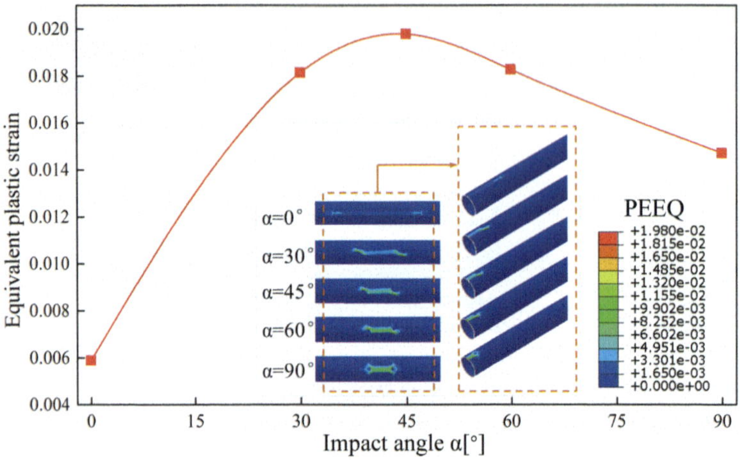

Fig. 3.16 Equivalent plastic strain and distributions of the submarine pipeline under different impact angles

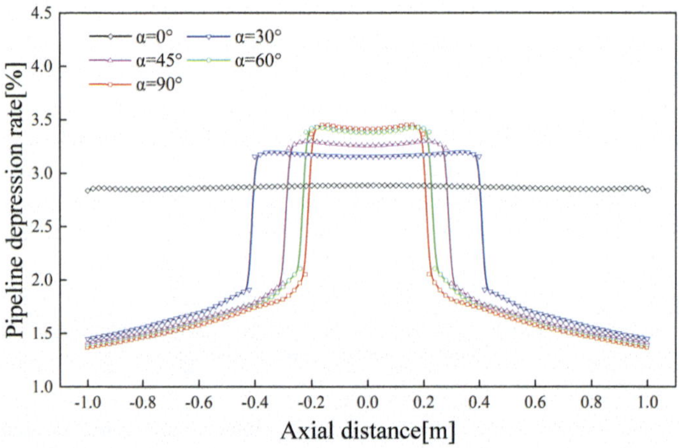

Fig. 3.17 Depression rate of the submarine pipeline under different impact angles

3.4.2.5 Pressure Difference

Submarine pipelines operate under varying pressure states due to the complex marine environment and unstable conditions. The inner and external pressure difference (P_D) significantly affects impact damage, influencing reliability and safety.

At an impact velocity of 5.482 m/s with constant impact energy, Fig. 3.19 shows equivalent plastic strain under different P_D values (− 0.9 MPa to 0.9 MPa). Plastic strain occurs only on the pipeline top, with little effect on its distribution. For P_D

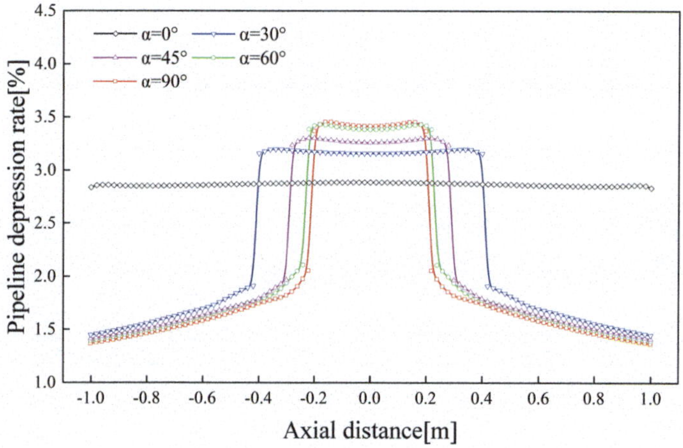

Fig. 3.18 Absorbed energy proportion and maximum impact force of the submarine pipeline under different impact angles

between − 0.3 MPa and 0.3 MPa, plastic strain decreases as P_D increases. However, when P_D exceeds 0.3 MPa, plastic strain increases with rising pressure difference.

Figure 3.20 shows depression rate curves along the axial distance under different pressure differences. At $P_D = -0.9$ MPa, the directly impacted area has the highest depression rate, but outside this zone, curves remain similar across pressure conditions. At $P_D = 0.9$ MPa, the depression rate is lowest due to internal pressure resistance. The maximum depression occurs at 0.2 m rather than the impact center (0 m). Overall, P_D has a minimal effect, with only a 0.64% difference between maximum and minimum depression rates.

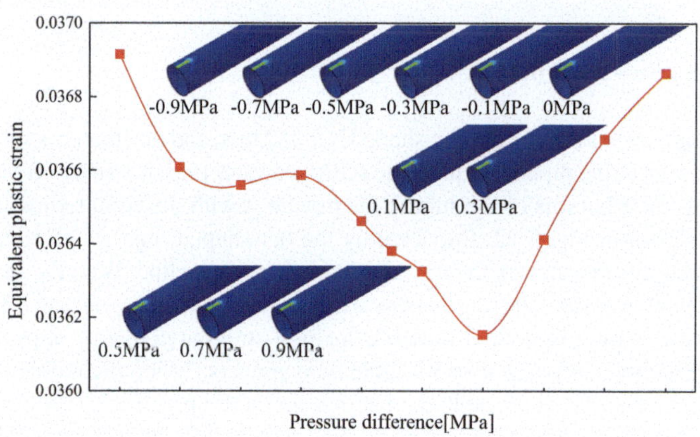

Fig. 3.19 Equivalent plastic strain and distributions of the submarine pipeline under different pressure differences

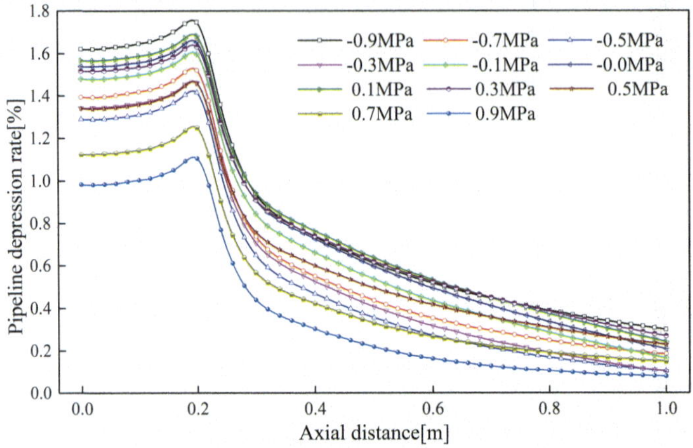

Fig. 3.20 Depression rate of the submarine pipeline under different pressure differences

The maximum impact forces and the absorbed energy proportions of the submarine pipeline impacted by the bar-shaped object under different pressure differences are shown in Fig. 3.21. With the increase of the pressure difference, the resisting effect of the submarine pipeline enhances. As the decrease of the deformability, the maximum impact force of the submarine pipeline increases with the rise of the pressure difference. When $-0.9\ \text{MPa} \leq P_D \leq 0.3\ \text{MPa}$, both the proportion of the energy absorbed by the submarine pipeline and its change rate gradually reduces with the increase of the pressure difference. When $P_D \geq 0.3$ MPa, the absorbed energy proportion shows an upward trend with the rise of P_D. Besides, the minimum absorbed energy proportion appears when the pressure difference equals 0.3 MPa, which is only 23.00%.

3.4.2.6 Radius-Thickness Ratio

When the impact velocity of the bar-shaped is 5.482 m/s, and the inner operation pressure of the submarine pipeline equals the static seawater pressure ($P_D = 0$ MPa), Von Mises stress distributions of the submarine pipelines with different radius-thickness ratios are shown in Fig. 3.22. Impacted by the bar-shaped object, deformation and stress concentration occur on the top of the submarine pipeline. With the increase of the radius-thickness ratio, the high-stress area on the top expands gradually. When the radius-thickness ratio is smaller than 60, the high-stress area mainly appears on the top of the pipeline, although no noticeable dent appears. When the radius-thickness ratio is bigger than 60, noticeable dent shows up on the pipeline's top; besides, the stress concentration occurs on both the left and right sides of the pipeline. Moreover, the stress concentration expands to the lower part of the submarine pipeline.

Figure 3.23 shows equivalent plastic strain and distributions of the submarine pipeline under different radius-thickness ratios. When the radius-thickness ratio of the

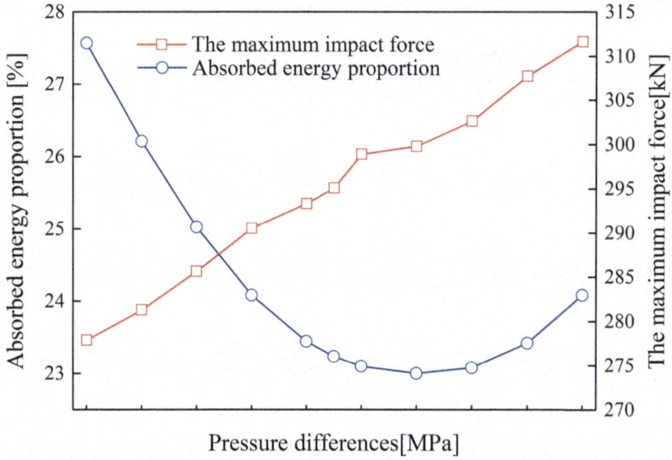

Fig. 3.21 Absorbed energy proportion and maximum impact force of the submarine pipeline under different pressure differences

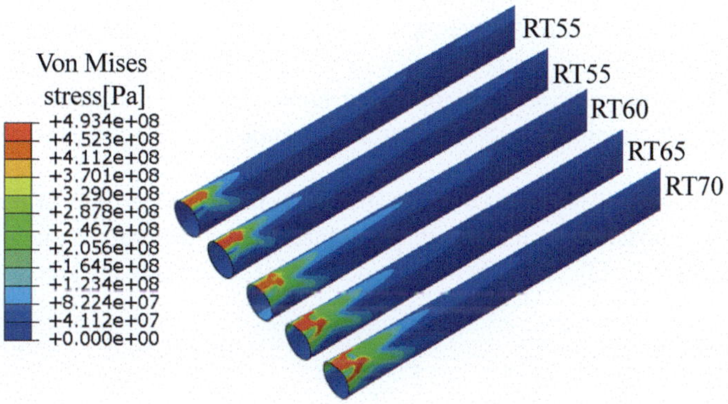

Fig. 3.22 Equivalent plastic strain and distributions of the submarine pipelines with different radius-thickness ratios impacted by the bar-shaped object

submarine pipeline is between 50 and 70, the plastic strain of the submarine pipeline increases with an unstable change rate as the radius-thickness ratio grows. When the radius-thickness ratio of the pipeline exceeds 60, plastic strain deduces gradually with the increase of the radius-thickness ratio. The larger the radius-thickness ratio is, the more severe plastic deformation on the top of the submarine pipeline becomes. Besides, plastic strain mainly distributes as "m" shape at the center and the edge of the dent caused by the impact. However, the plastic strain area changes less along the axial direction with the change of the radius-thickness ratio.

Figure 3.24 illustrates the impact of the radius-thickness ratio on submarine pipeline deformation. With a constant pipeline radius, increasing the ratio reduces

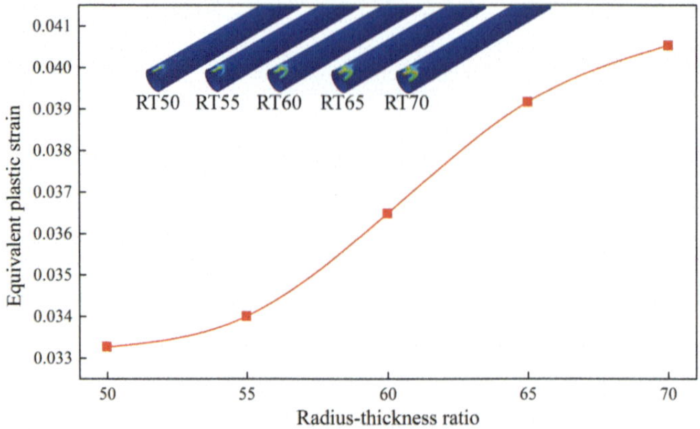

Fig. 3.23 Equivalent plastic strain and distributions of the submarine pipeline under different radius-thickness ratios

wall thickness, weakening structural resistance and significantly increasing the depression rate. At a ratio of 30, the maximum depression rate is 0.907%, but it rises to 8.002% at 70. Proper selection of this ratio is critical for balancing structural reliability and manufacturing costs. As wall thickness decreases, visible dents form on the top, and outward expansion appears on both sides. When the ratio is below 50, the maximum depression occurs at 0.2 m, but for ratios above 50, it shifts slightly to 0.198 m.

The maximum impact forces and the absorbed energy proportions of the submarine pipeline impacted by the bar-shaped object under different radius-thickness ratios are shown in Fig. 3.25. With the increase of the radius-thickness ratio, the

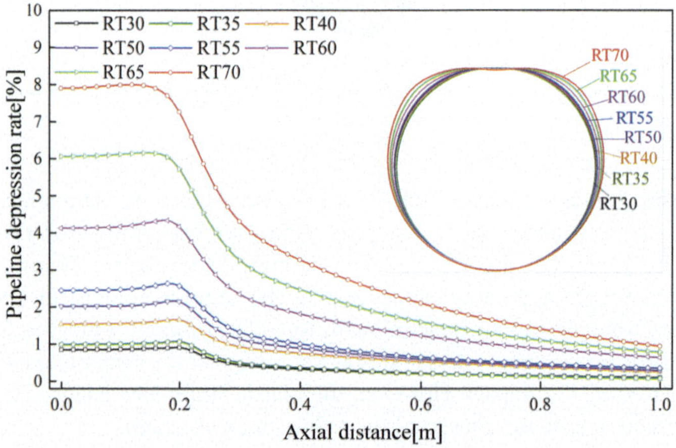

Fig. 3.24 Depression rate of the submarine pipeline under radius-thickness ratios

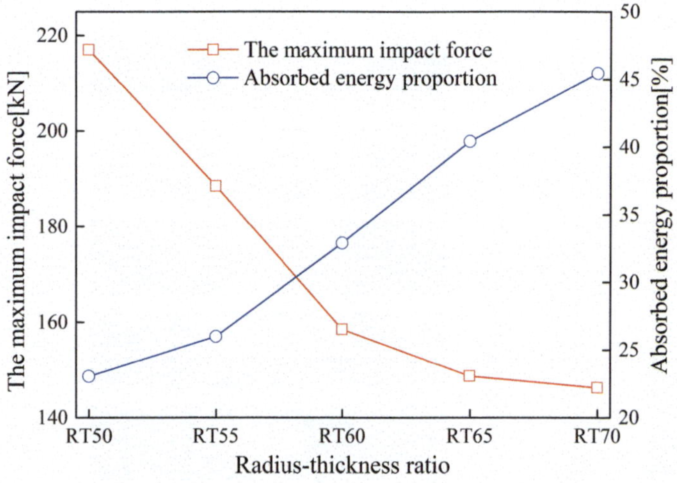

Fig. 3.25 Absorbed energy proportion and maximum impact force of the submarine pipeline under different radius-thickness ratios

maximum impact force and its change rate both decline. On the contrary, as the radius-thickness ratio increases, the proportion of the energy absorbed by the submarine pipeline shows a nonlinear upward trend.

3.4.2.7 Inclined Impact Angle

With the effects of external factors, such as the ocean current, falling objects may impact submarine pipelines with an inclined angle. Figure 3.26 shows the schematic diagram of an inclined impact. The inclined impact angle β is defined as the angle between the pipeline axis and the impact velocity of the bar-shaped object. In this section, the initial impact velocity is 6 m/s, and the internal and external pressure are both 1 MPa.

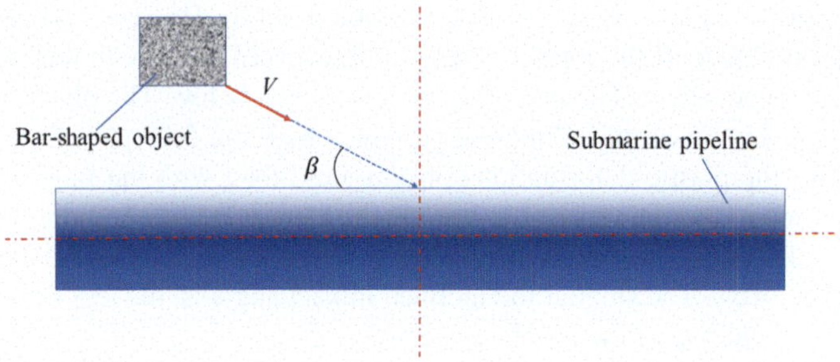

Fig. 3.26 Schematic diagram of inclined impact

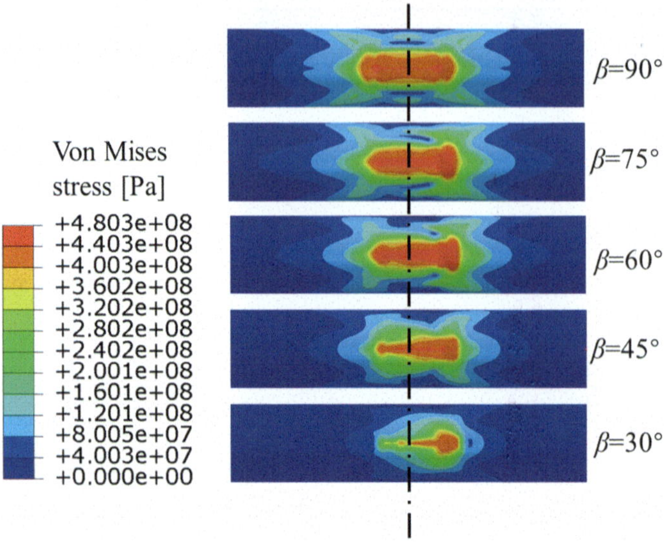

Fig. 3.27 Equivalent plastic strain and distributions of the submarine pipeline impacted by the bar-shaped object with the inclined impact angle ranges from 90° to 30°

The Von Mises stress distributions of the submarine pipeline impacted by the bar-shaped object with different inclined impact angle are shown in Fig. 3.27. When $\beta = 90°$, the bar-shaped object impact transversely impacts the pipeline. The Von Mises distribution on the top of the pipeline is symmetric along the dashed line. With the decease of the inclined impact angle, the dissymmetry of the stress concentration becomes more and more noticeable. When $\beta = 30°$, most of the stress concentration located on the right side of the dashed line. When the inclined impact angle equals 75° and 60°, the high-stress areas have similar distributions. If the inclined impact angle ranges from 75° to 30°, the area of the stress concentration becomes smaller with the decrease of the inclined impact angle.

As shown in Fig. 3.28, when the inclined impact angel is smaller than 60°, the equivalent plastic strain increases with the rise of the inclined impact angle. The maximum equivalent plastic strain appears when $\beta = 60°$. However, when $\beta > 60°$, the equivalent plastic of the pipeline shows a downward trend with the increase of the inclined impact angle. The rate of descent is relatively low. The plastic deformation occurs at the top of the submarine pipeline. With the decrease of the inclined impact angle, the plastic deformation concentrates along with the impact velocity. The distribution of the plastic strain on the pipeline subjected to a 75° inclined impact is similar to that of the pipeline subjected to a 60° inclined impact. The distribution seems like an arrow. The smaller the inclined impact angle is, the less severe plastic deformation on the top of the pipeline becomes.

With the inclined impact caused by the bar-shaped object, submarine pipelines will be dented in different degrees. The depression rates of the submarine pipeline along with the axis under varying inclined angles are shown in Fig. 3.29. When the

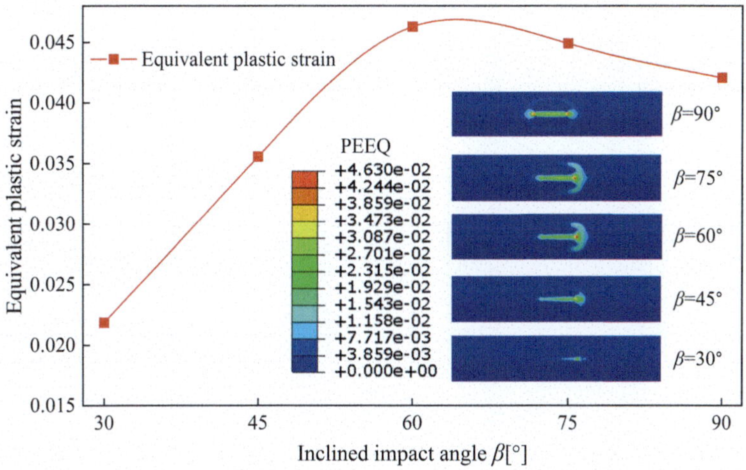

Fig. 3.28 Equivalent plastic strain and distributions of the submarine pipeline under different inclined impact angles

inclined angle is 90°, the bar-shaped object vertically impacts the submarine pipeline, and the bottom of the dent is relatively flat. When the bar-shaped object inclines, the deformation of the pipeline changes. The bottom of the dent will also slope. When the inclined impact angle ranges from 75° to 30°, the depression rate on the right of the dashed line is noticeably higher than that on the left side, and the depression rates decline with the decrease of the inclined angle.

The absorbed energy proportions of the submarine pipeline impacted by the bar-shaped object with different inclined impact angles are shown in Fig. 3.30. With the increase of the inclined impact angle, the energy absorbed by the submarine

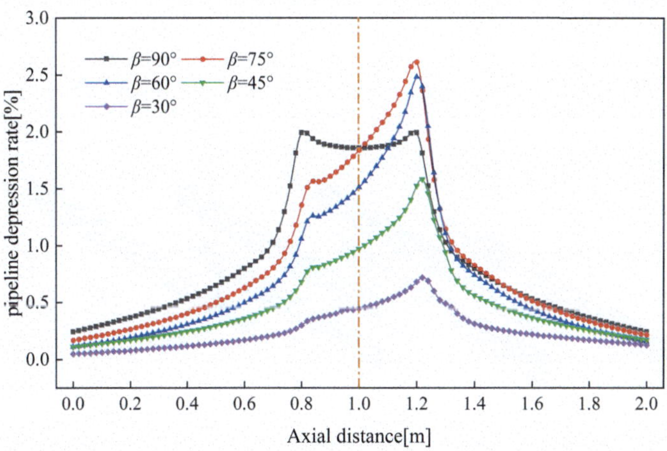

Fig. 3.29 Depression rate of the submarine pipeline under different inclined impact angles

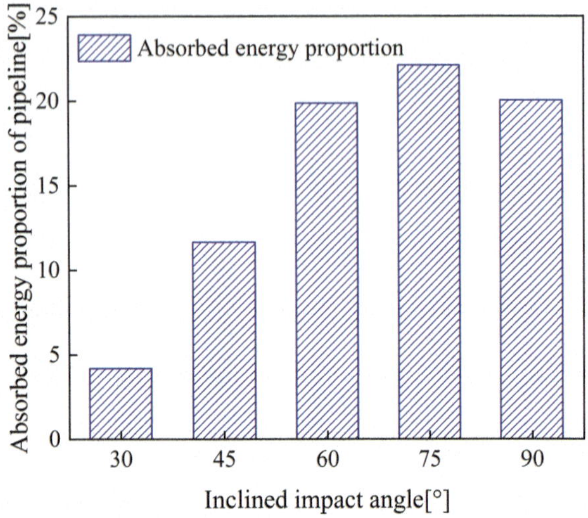

Fig. 3.30 Absorbed energy proportion of the submarine pipeline under different inclined impact angles

pipeline first increases and then decreases. When $\beta = 75°$, the proportion of the energy absorbed by the pipeline is the maximum.

3.4.3 *Sensitivity Analysis for Lump-Shaped Object Impact*

3.4.3.1 Impact Velocity

Figure 3.31 shows Von Mises stress distributions under different impact velocities, which are influenced by external factors like ocean currents and initial velocity. High-stress areas appear at the impact point on the pipeline top, intensifying with higher velocities. As velocity increases, stress spreads laterally and axially. Above 10 m/s, a noticeable dent forms, with peak stress at the indentation center and lower stress at its edges. At 14 m/s, stress concentration extends to the pipeline bottom.

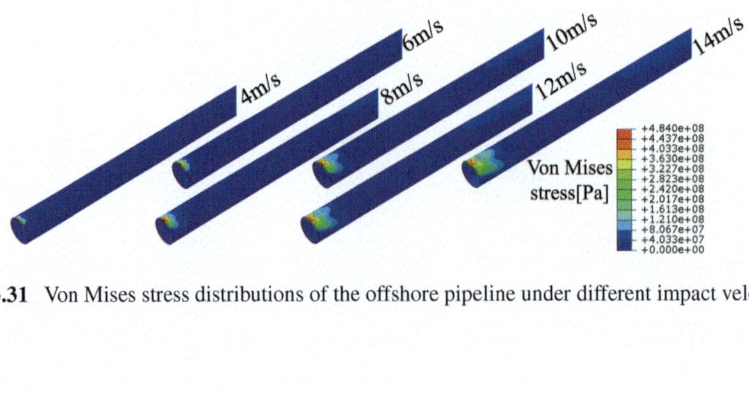

Fig. 3.31 Von Mises stress distributions of the offshore pipeline under different impact velocities

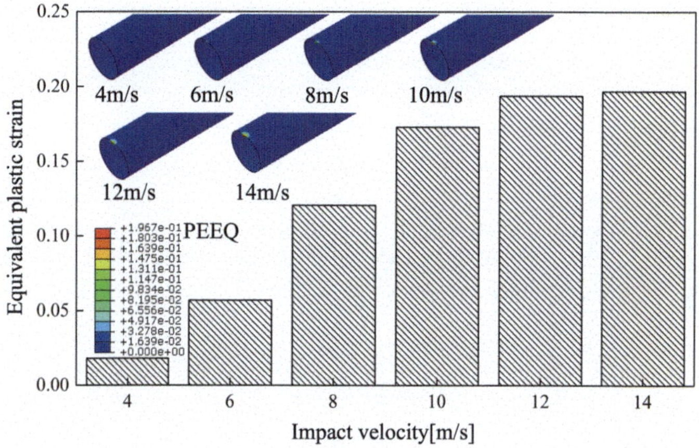

Fig. 3.32 Equivalent plastic strain and distributions of the offshore pipeline under different impact velocities

Figure 3.32 presents plastic strain distributions under different impact velocities. Below 8 m/s, plastic deformation is minimal and confined to the impact area. As velocity increases, the deformed area expands, covering the entire dent above 10 m/s. Equivalent plastic strain rises nonlinearly, increasing rapidly below 10 m/s but slowing significantly beyond 12 m/s.

The depression ratio is introduced to quantify the deformation of the offshore pipeline caused by impact in this paper. The depression ratio is defined by Eq. 3.12.

$$\delta_d = \frac{(D_o - D_c)}{D_o} 100\% \tag{3.12}$$

where δ_d is the depression ratio of the offshore pipeline, D_o is the original diameter of the pipeline before impact (m), D_c is the current diameter of the pipeline after impact (m).

The depression ratio, along with the distance of the impact area (0–1 m) and cross-sectional deformations of the offshore pipeline under different impact velocities are shown in Fig. 3.33 When the impact velocity of the lump-shaped object is lower than 4 m/s, no visible deformation occurred in the cross-section of the pipeline, and the depression ratio is quite small, less than 1%. The cusp-shape dent gradually collapses to the center of the cross-section with the increase of the impact velocity. Besides, the convex height of the dented edge gradually increases due to the deformation and the squeezing action. The higher the impact velocity is, the greater the impact energy of the falling object has, which means more energy needs to be absorbed by the offshore pipeline. The depression ratio of the offshore pipeline increases with the increase of impact velocity, as well as the deformation area along the axial direction.

Figure 3.34 shows the maximum depression ratio of the offshore pipeline subjected to impact damage as a function of the impact velocity of the lump-shaped

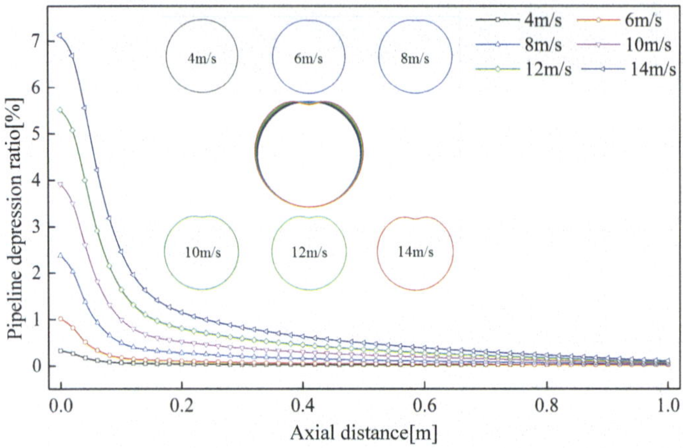

Fig. 3.33 Depression ratio and cross-sectional deformation of the offshore pipeline under different impact velocities

object. An apparent linear relationship can be observed between the maximum depression ratio and the impact velocity. According to DNV-RP-F107 [1], the critical depression ratio of the offshore pipeline if 5%. If the maximum depression ratio of the dented pipeline exceeds 5%, maintenance action is required. Through the fitted curve, it can be easily estimated that the critical impact velocity that induces a 5% depression ratio is around 11.1 m/s. Thus, the critical impact energy of the lump-shaped object for failure limit state is about 9733.3 J.

Fig. 3.34 The maximum depression ratio as a function of the impact velocity of the lump-shaped object

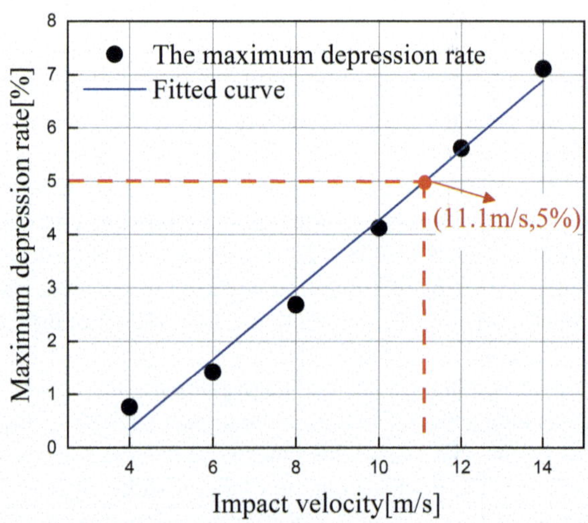

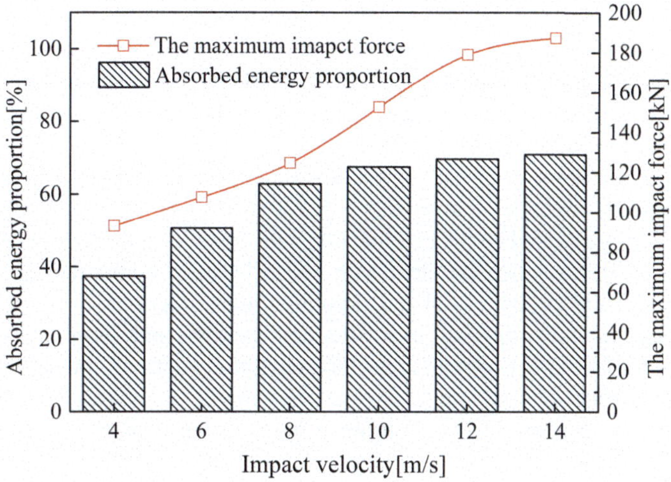

Fig. 3.35 Absorbed energy proportion and the maximum impact force of the offshore pipeline under different impact velocities

Figure 3.35 shows the absorbed energy proportion and the maximum impact force of the offshore pipeline impacted by the lump-shaped object with different velocities. Through deformation, tubular structures could absorb impact energy caused by falling objects. The absorbed energy proportion of the deformed offshore pipeline is defined as Eq. 3.13.

$$P_{abs} = \frac{E_I}{E_A} \tag{3.13}$$

where P_{abs} is the absorbed energy proportion of the offshore pipeline after impact, E_I is the internal energy of the deformed pipeline at the final stage (J), E_A is the total energy of the system (J).

When the impact velocity ranges from 4 to 10 m/s, both the maximum impact force and the absorbed energy proportion of the offshore pipeline increased rapidly with the increase of the impact velocity. However, when the impact velocity is higher than 12 m/s, the proportion of the energy absorbed by the offshore pipeline almost remains stable, owing to the limit of the pipeline's absorptive capacity of the energy. The change rate of the maximum impact force of the offshore pipeline decreases gradually after the impact velocity is higher than 10 m/s.

3.4.3.2 Impact End-Shape

In practice, different kinds of falling objects might be caused by human activities on the sea, such as containers, drills, anchors, etc. In some situations, the submarine pipeline might be impacted by a falling object with sharp corners. It is reasonable to idealize the complex shape of the falling object to make the research feasible. As shown in Fig. 3.36, four typical shapes of the impact end are selected as samples to be analyzed, including the hexahedral end, cylindrical end, conical end, and spherical end. These four different lump-shaped objects have the same projected area of the object in the flow direction and volume.

Figure 3.37 presents Von Mises stress distributions for lump-shaped objects with different impact-end shapes at a velocity of 9.228 m/s. Due to the small impact area, high stress is concentrated on the pipeline top, with no stress at the bottom. A conical impact end creates a noticeable dent with an "m"-shaped high-stress area, which expands significantly in both axial and circumferential directions. In contrast, hexahedral and cylindrical ends result in smaller deformations and lower stress concentrations due to their larger contact area, enhancing the pipeline's impact resistance. Their stress distributions are similar. A spherical impact end causes a shallow dent, with high stress appearing at the bottom of the dent but without significant axial expansion.

Figure 3.38 presents plastic strain distributions and equivalent plastic strain for different impact-end shapes. For hexahedral and cylindrical (flat-end) impacts, only minor plastic strain occurs, mainly at the edge of the impact area, forming an elliptical pattern rather than centering on the impact point. A conical impact end causes the most severe plastic strain, highly concentrated at the dent center. A spherical impact end results in greater strain than cylindrical and hexahedral ends but slightly less than the conical end. Among all shapes, the conical impact end causes the most severe damage and the largest plastic strain area.

Lump-shaped objects with different impact-end shapes create varying dent shapes, sizes, and depths on offshore pipelines. Figure 3.39 shows the depression ratio and cross-sectional deformation under different impact ends. A conical impact end causes the most severe deformation, with a significantly higher depression ratio than the other cases. Its cusp-shaped dent extends toward the cross-section center. In contrast,

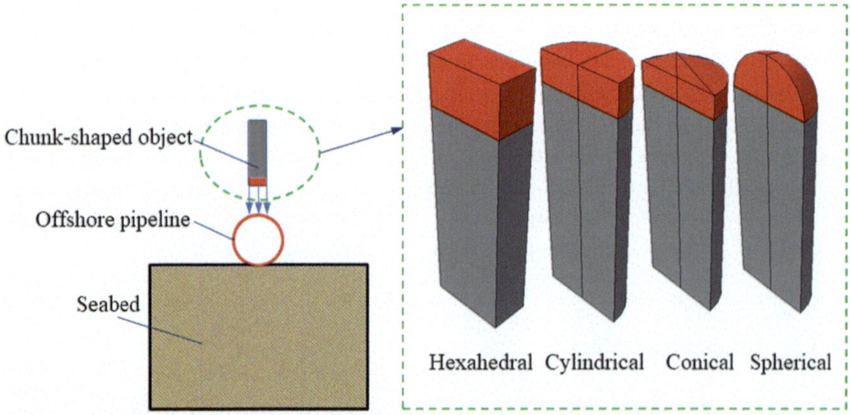

Fig. 3.36 Schematic diagram of the lump-shaped falling object with different impact-end shapes

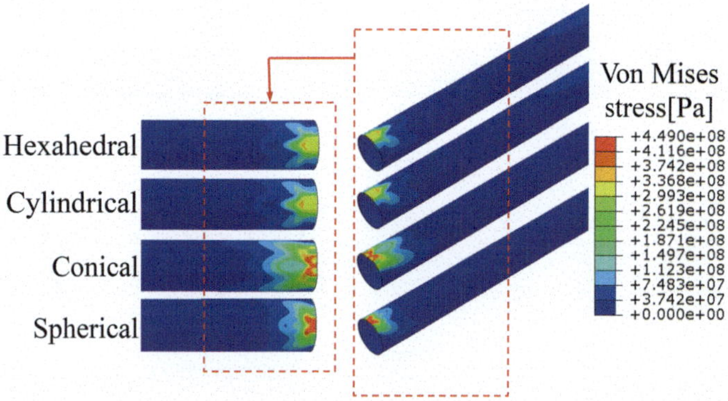

Fig. 3.37 Von Mises stress distributions of the offshore pipeline under different impact-end shapes

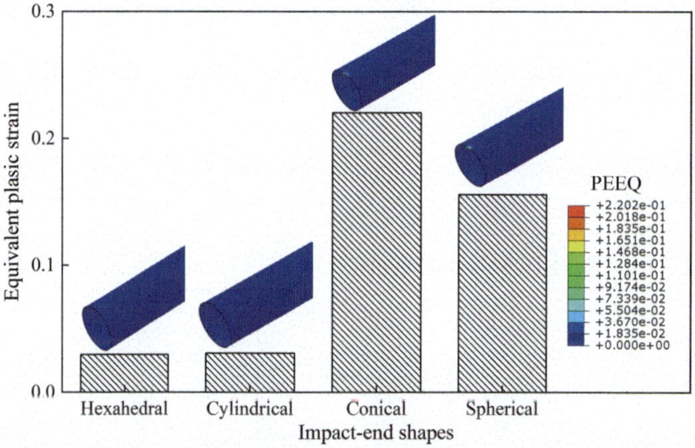

Fig. 3.38 Equivalent plastic strain and distributions of the offshore pipeline under different impact-end shapes

cylindrical and hexahedral ends produce a more uniform dent bottom with minimal variation in depression ratio. Regardless of shape, the depression ratio decreases with axial distance. Convex ends cause more severe local and global damage than flat ends, with the maximum depression ratio for a cylindrical end being only 12.654% of that for a conical end.

Figure 3.40 presents the absorbed energy proportions and maximum impact force for different impact-end shapes. The cylindrical impact end generates the highest impact force, with energy absorption similar to the hexahedral end. The conical end results in the lowest impact force at 65.56 kN, only 32.95% of that from the cylindrical end. The spherical end leads to the highest energy absorption, with the conical end showing a similar absorption proportion.

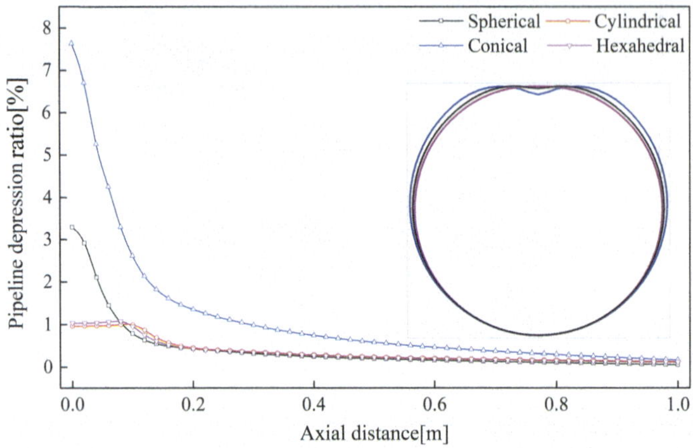

Fig. 3.39 Depression ratio and cross-sectional deformation of the offshore pipeline under different impact-end shapes

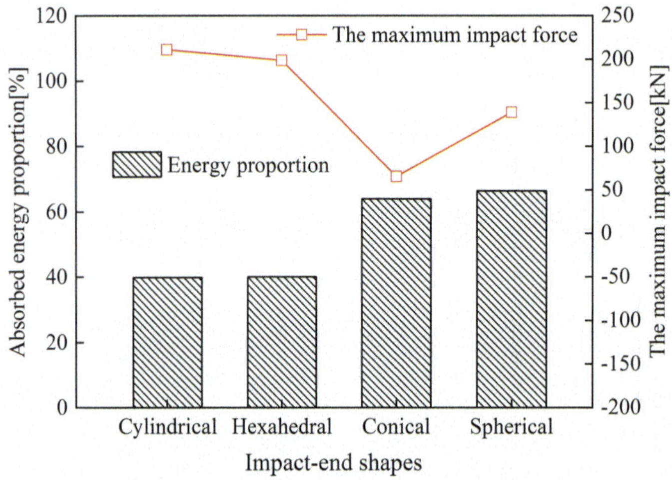

Fig. 3.40 Absorbed energy proportion and the maximum impact force of the offshore pipeline under different impact-end shapes

3.4.3.3 Lump-Shaped Object Size

Lump-shaped objects, which have the same volume, different sphere-shaped impact ends, are selected to be analyzed. The geometrical parameters of the lump-shaped object with a sphere-shaped impact end are shown in Table 3.2.

Figure 3.41 shows Von Mises stress distributions for offshore pipelines impacted by lump-shaped objects of different sizes but with equal impact energy. A high-stress zone consistently appears on the pipeline top, with no stress concentration at

Table 3.2 Geometrical parameters of the lump-shaped object with sphere-shaped impact end

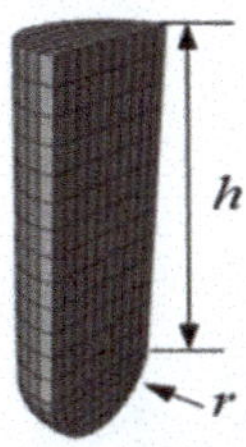

Size number	Radius r(m)	Height h (m)
Size1	2×10^{-1}	8.333×10^{-3}
Size2	1.6×10^{-1}	1.147×10^{-1}
Size3	1.4×10^{-1}	1.958×10^{-1}
Size4	1.2×10^{-1}	3.314×10^{-1}
Size5	1×10^{-1}	5×10^{-1}
Size6	8×10^{-2}	8.321×10^{-1}
Size7	6×10^{-2}	1.534

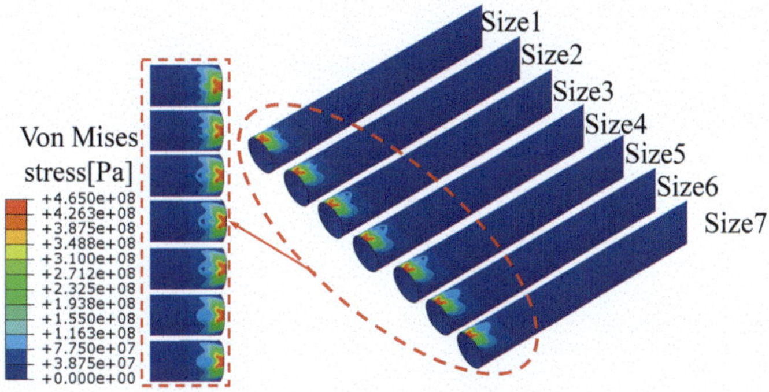

Fig. 3.41 Von Mises stress distributions of the offshore pipeline under different sizes of the lump-shaped object

the bottom due to the limited impact energy. For size 1 and size 2, stress distribution remains relatively uniform. When the size exceeds 2, stress extends axially, forming a "w" shape, and expands circumferentially as the size increases.

The equivalent plastic strain and plastic strain distributions of the offshore pipeline impacted by lump-shaped objects with different sizes are shown in Fig. 3.42. With the rise of the size number, the area of the spherical impact end becomes smaller, which leads to the concentration of the plastic deformation in the impact dent. The most severe plastic strain occurs on the top of the offshore pipeline when the size number of the falling object is 7. With the increase of the size number, the equivalent plastic strain of the offshore pipeline at the final state shows a general upward trend, although the equivalent plastic strain value of size 4 is higher than that of size 5.

Figure 3.43 shows the depression ratio and cross-sectional deformation of the offshore pipeline after impact from lump-shaped objects of different sizes. The depression ratio at the dent center is affected by the object size, with larger sizes concentrating the impact force, thus increasing the depression ratio. However, beyond 0.05 m axial distance, the depression ratios become similar. The global deformation and cross-sectional deformation of the pipeline are only slightly influenced by the object size.

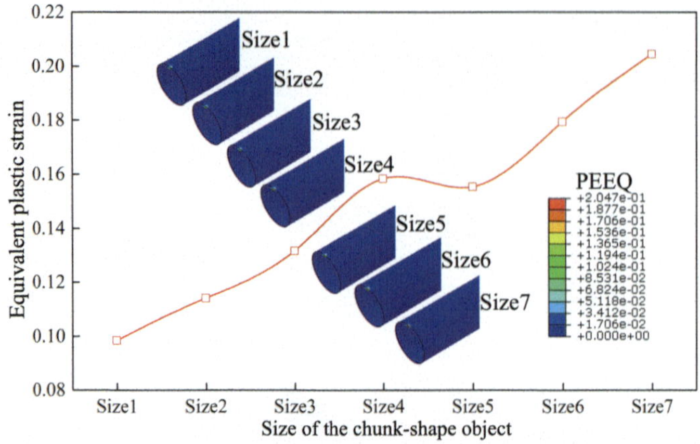

Fig. 3.42 Equivalent plastic strain and distributions of the offshore pipeline under different sizes of the lump-shaped object

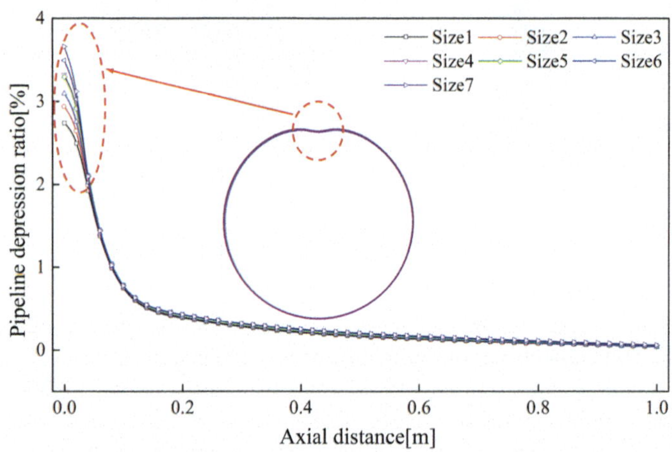

Fig. 3.43 Depression ratio and cross-sectional deformation of the offshore pipeline under different sizes of the lump-shaped object

Figure 3.44 shows the absorbed energy proportion and maximum impact force of the offshore pipeline under different lump-shaped object sizes. The maximum impact force decreases nonlinearly with the increase in size number. The proportion of energy absorbed by the pipeline increases with size, but the rate of change is not constant, with the minimum rate occurring between Size 4 and Size 5. Thus, a narrow impact end causes more severe damage than a broad one, given the same mass, impact velocity, and impact end shape.

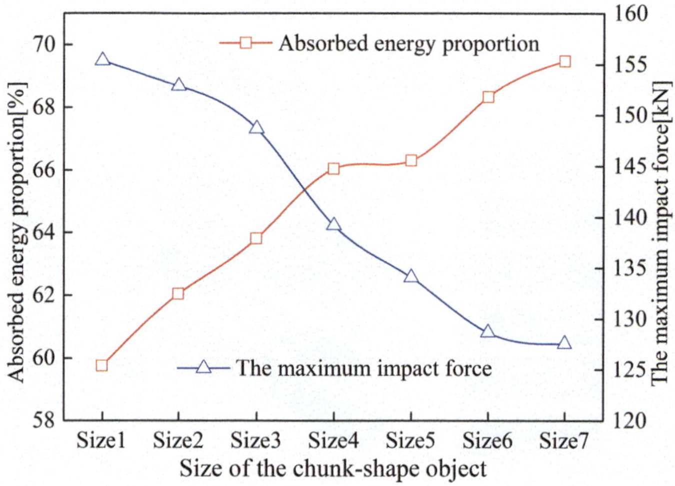

Fig. 3.44 Absorbed energy proportion and the maximum impact force of the offshore pipeline under different lump-shaped object sizes

3.4.3.4 Eccentric Distance

It is necessary to take the eccentric distance between the pipeline and the falling object into consideration. The schematic diagram of the offshore subjected to an eccentric impact is shown in Fig. 3.45 *l* is the distance between the falling object and the offshore pipeline (m).

Fig. 3.45 Schematic diagram of an eccentric impact

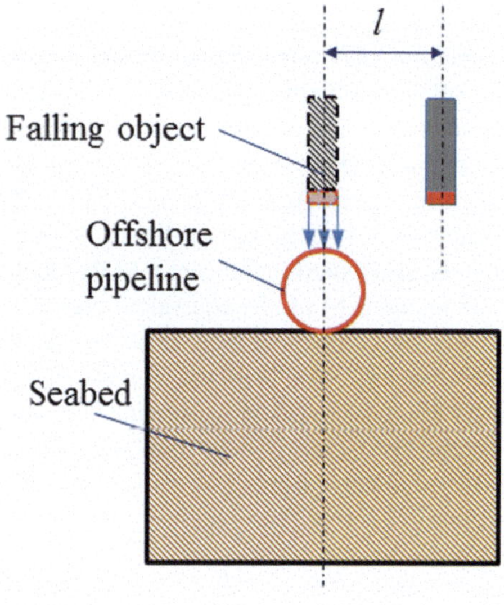

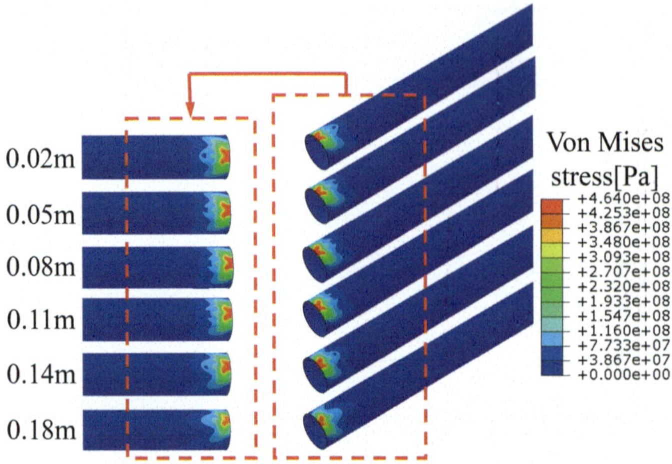

Fig. 3.46 Von Mises stress distributions of the offshore pipeline under different eccentric distances

Von Mises stress distributions of the pipeline impacted by the lump-shaped object under different eccentric distances are shown in Fig. 3.46 The bigger the eccentric distance is, the larger the deviated distance between the high-stress area and the pipeline top becomes. Besides, the slightest stress concentration of which the shape is different from the other conditions appears on the left of the pipeline top when the eccentric distance is 0.18 m. When the eccentric distance ranges from 0.02 m to 0.11 m, the shapes of the highly stressed zones are very similar while the positions are different; moreover, no distinct stress concentration is observed on the bottom of the pipeline. With the increase of the eccentric distance, the area of the high stress decreases.

Figure 3.47 shows the equivalent plastic strain and plastic strain distributions of the offshore pipeline impacted by the lump-shaped object with different eccentric distance. While the minimum equivalent plastic strain of the offshore pipeline appears when the eccentric distance is 0.08 m, the maximum equivalent plastic strain appears when the eccentric distance is 0.14 m. When the eccentric distance is 0.14 m, a distinct dent is observed on the right side of the pipeline, and the plastic strain is mainly distributed in the center of the dent. When the eccentric distance is between 0.08 m and 0.14 m, the equivalent plastic strain increases nonlinearly.

The curve of the energy absorbed by the offshore pipeline at the final stage is shown in Fig. 3.48. With the increase of the eccentric distance, the absorbed energy proportion of the offshore pipeline declines gradually; besides, the rate of change of the absorbed energy proportion significantly grows up when the eccentric distance is larger than 0.11 m.

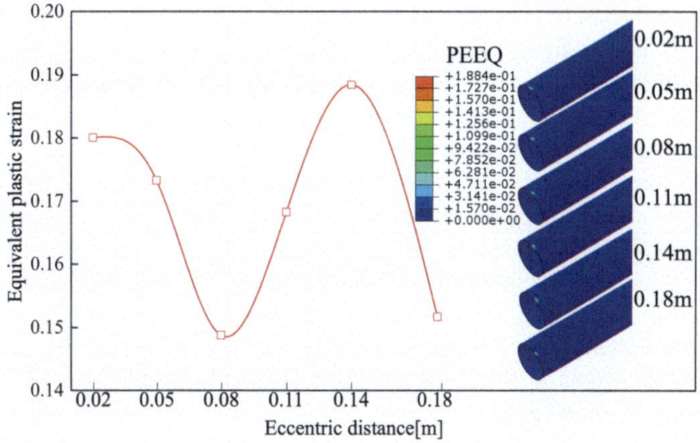

Fig. 3.47 Equivalent plastic strain and distributions of the offshore pipeline under different eccentric distances

Fig. 3.48 Absorbed energy proportion of the offshore pipeline under different eccentric distances

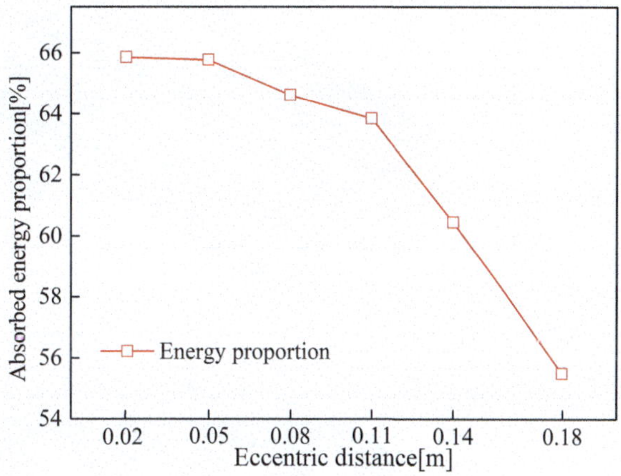

3.4.3.5 Pressure Difference

The pressure difference of the offshore pipeline is defined as the following function:

$$P_D = P_I - P_O \tag{3.14}$$

where P_D is the pressure difference (MPa); P_I is the internal operating pressure of the pipeline (MPa); P_O is the outer pressure in the external environment (static seawater pressure) (MPa).

Von Mises stress distributions of the offshore pipeline under different pressure differences are shown in Fig. 3.49. When the range of the pressure difference of the offshore pipeline is between − 0.9 and − 0.5 MPa, the stress concentration on the top of the pipeline not only expands axially but also extends circumferentially to the lateral side of the pipeline. When the P_D is from − 0.9 to 0.9 MPa, the high-stress area shows up in the impact area on the offshore pipeline's top, and the high-stress regions are similar in shape under different pressure difference. When the P_D equals 0.7 MPa, stress concentration occurs on the left and right sides of the end of the offshore pipeline.

Figure 3.50 shows the equivalent plastic strain and plastic strain distributions of the offshore pipeline impacted by the lump-shaped object under different pressure differences. The apparent plastic strain occurs on the impact area, and the most severe plastic strain is observed at the edge of the pipeline's end; in addition, the plastic strain area expands with the increase of the pressure difference on a very small scale. With the rise of the pressure difference, the equivalent plastic strain grows gradually, but the magnitude of the change is tiny.

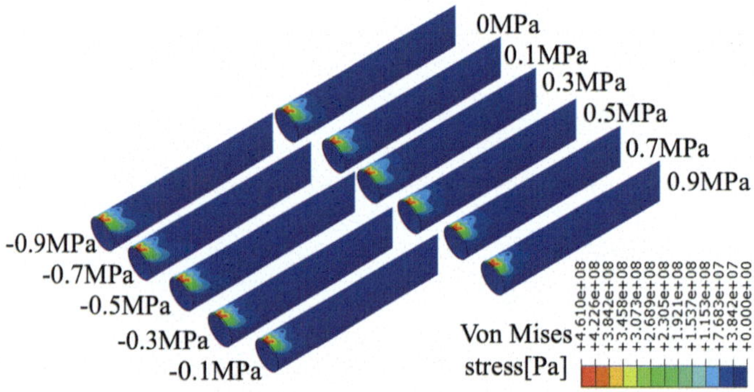

Fig. 3.49 Von Mises stress distribution of the offshore pipeline under different pressure differences

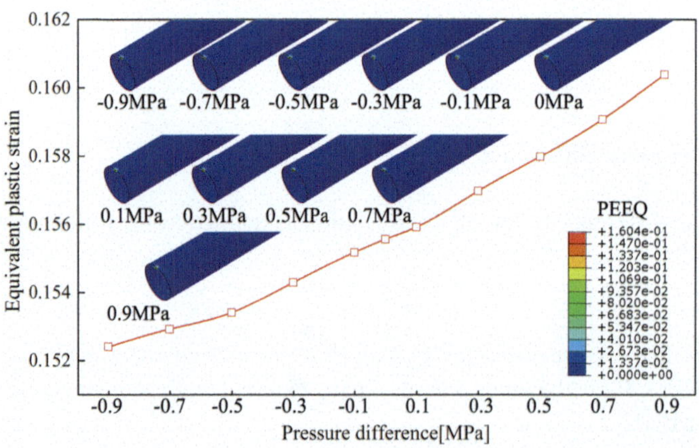

Fig. 3.50 Equivalent plastic strain and distributions of the offshore pipeline under different pressure differences

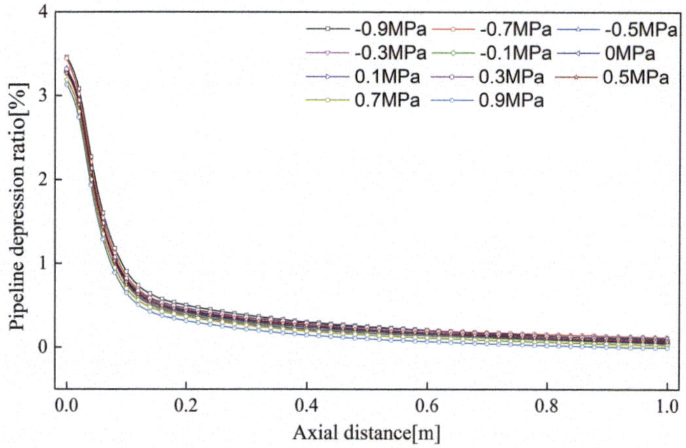

Fig. 3.51 Depression ratio of the offshore pipeline under different pressure differences

Figure 3.51 shows the depression ratio of the offshore pipeline along the axial distance under different pressure differences. The depression ratio trends are similar across pressure differences from − 0.9 to 0.9 MPa. Due to internal pressure resistance, the lowest depression ratio occurs at 0.9 MPa, while the highest ratio is observed when the pressure difference is between − 0.9 and − 0.7 MPa. Overall, the pressure difference has a slight effect, with the minimum depression ratio at 90.45% of the maximum.

Figure 3.52 shows the absorbed energy proportions and maximum impact force of the offshore pipeline impacted by the lump-shaped object under different pressure differences. The maximum impact force increases with pressure difference, although the force at − 0.5 MPa is slightly higher than at − 0.3 MPa. For pressure differences between − 0.9 and 0 MPa, the absorbed energy proportion and its rate of change decrease with increasing pressure difference. In contrast, for pressures between 0 and 0.9 MPa, the absorbed energy proportion increases as the pressure difference rises.

3.4.3.6 Radius-Thickness Ratio

Figure 3.53 shows the Von Mises stress distributions of offshore pipelines with different radius-thickness ratios. With a constant radius of 0.508 m, as the radius-thickness ratio increases, the wall thickness decreases, weakening the pipeline's resistance capacity. This causes the high-stress area on the top to grow and expand along the axial direction. For ratios below 50, the high-stress area is at the edge of the pipeline, with no stress at the bottom. Above a ratio of 50, the top of the pipeline dents noticeably, and the high-stress area forms an "m" shape.

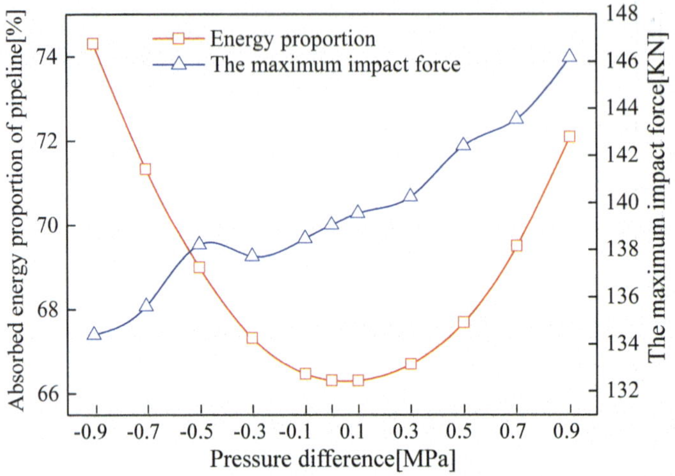

Fig. 3.52 Absorbed energy proportion and the maximum impact force of the offshore pipeline under different pressure differences

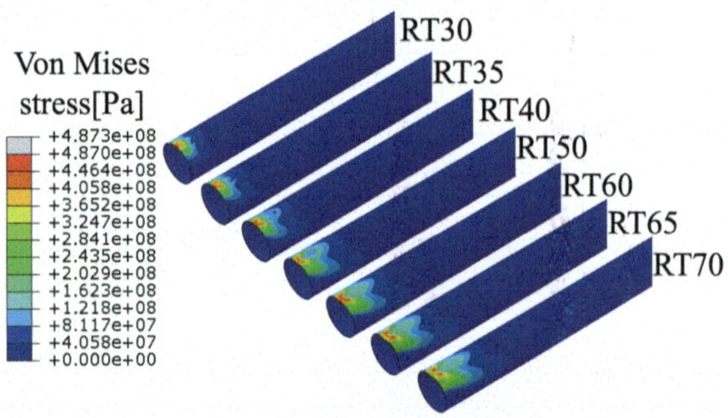

Fig. 3.53 Von Mises stress distributions of the offshore pipeline under different radius-thickness ratios

Equivalent plastic strain and plastic strain distributions of the offshore pipeline impacted by the lump-shaped object under different Radius-thickness ratios are shown in Fig. 3.54. With the increase of the radius-thickness ratio, the plastic strain area caused by the lump-shaped object gradually expands. No plastic strain appears on neither the bottom nor the left and right sides of the impacted pipeline. When the radius-thickness ratio is within the 30–40, the equivalent plastic strain shows an increasing trend as the radius-thickness ratio increases. However, the equivalent plastic strain decreases with the rise of the radius-thickness ratio when the radius-thickness is bigger than 50. Besides, the maximum equivalent plastic strain appears when the radius-thickness is 40.

Figure 3.55 shows the cross-sectional deformations and depression ratios of the offshore pipeline with different radius-thickness ratios impacted by a lump-shaped object. A smaller wall thickness reduces the pipeline's structural stiffness, making it

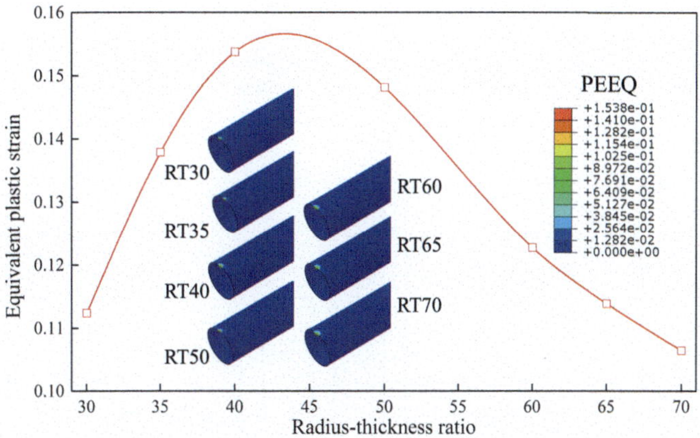

Fig. 3.54 Equivalent plastic strain and distributions of the offshore pipeline under different radius-thickness ratios

more susceptible to deformation. As the radius-thickness ratio increases, the depression ratio in the impact area rises, with the depression ratio decreasing significantly along the axial distance. At a ratio of 70, the deformation extends widely along the axis, with the depression ratio dropping below 1% beyond 0.36 m. Larger radius-thickness ratios lead to greater cross-sectional deformation and deeper impact pits.

The maximum impact force during the impact process and the absorbed energy proportion at the final state of the offshore pipeline impacted by the lump-shaped object under different radius-thickness ratios are shown in Fig. 3.56. As the increase of the radius-thickness ratio, the proportion of energy absorbed by the offshore

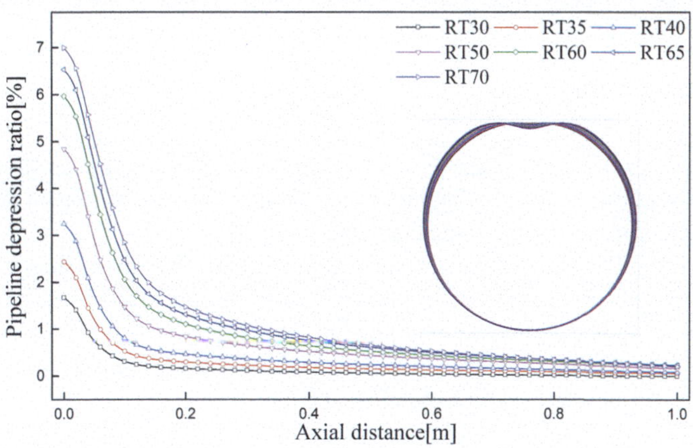

Fig. 3.55 Depression ratio and cross-sectional deformation of the offshore pipeline under different radius-thickness ratios

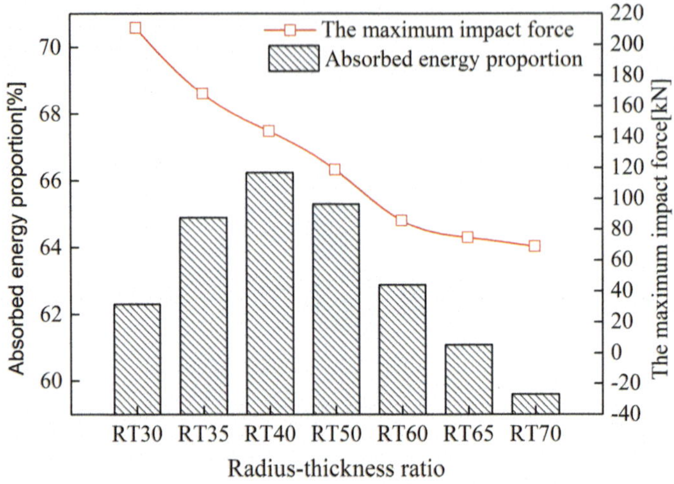

Fig. 3.56 Absorbed energy proportion and the maximum impact force of the offshore pipeline under different radius-thickness ratios

pipeline gradually increases to 60%, when the radius-thickness ratio increases from 30 to 40. However, after RT40, the absorbed energy proportion of the pipeline decreases gradually with the growth of the radius-thickness. Moreover, the maximum impact force of the offshore pipeline decreases nonlinearly with the increase of the radius-thickness; and when the radius-thickness is 70, the energy absorbed by the impacted offshore pipeline takes the least portion of the total energy. With the increase of the radius thickness ratio, i.e., the decrease of the wall thickness, the volume of the pipeline becomes smaller, which leads to the weakening of the ability to absorb energy.

3.5 Conclusions

This chapter examines the damage mechanisms of offshore pipelines subjected to impacts from falling objects, using numerical simulations for bar-shaped and lump-shaped object impacts. Key findings include:

Impact Type: Bar-shaped objects cause more damage than lump-shaped objects, with the seabed absorbing most of the impact energy. This underscores the importance of considering seabed composition in pipeline design.

Seabed Composition: Rocky seabeds lead to severe plastic deformation, suggesting pipeline routes should avoid such areas.

Impact Velocity: Higher velocities increase stress concentration, plastic deformation, and impact force, highlighting the need for robust materials and construction methods.

Radius-Thickness Ratio: A larger ratio increases damage and energy absorption, emphasizing the importance of optimizing this ratio for pipeline resilience.

Object Shape and Impact Angle: Tri-Prism ends and impact angles of 60°-75° cause the most severe damage.

Lump-shaped Objects: Seabed flexibility mitigates deformation, but higher velocities cause more severe damage, with critical failure at 9733.338 J.

Impact End Shape and Size: Convex ends lead to greater deformation. The eccentric distance between the object and the pipeline also affects damage severity.

Pressure Differences: Higher pressure differences reduce dent depth but increase plastic strain and impact force.

In conclusion, offshore pipeline integrity is influenced by seabed characteristics, geometry, impact dynamics, and environmental pressures. This research provides insights for safer, more reliable pipeline design and protection.

This chapter is adapted from the articles 'Numerical Study on Mechanism Responses of Submarine Pipeline Impacted by Bar-Shaped Falling Object,' originally published in *Journal of Pipeline Systems Engineering and Practice* (2022), DOI: https://doi.org/10.1061/(ASCE)PS.1949-1204.0000505, and 'Numerical Investigation of The Damage Mechanism of Offshore Pipeline Impacted by The Lump-Shaped Falling Object,' originally published in *Journal of Pressure Vessel Technology* (2021), DOI: https://doi.org/10.1115/1.4049146. Reproduced with permission from ASCE and ASME.

References

1. DNV (Det Norske Veritas) (2010) Risk assessment of pipeline protection. DNV-RP-F107. DNV, Oslo, Norway
2. Bai Y, Pedersen PT (1993) Elastic-plastic behaviour of offshore steel structures under impact loads. Int J Impact Eng 13(1):99–115
3. Ghaednia H, Das Sreekanta WR, Richard K (2017) Dependence of burst strength on crack length of a pipe with a dent-crack defect
4. Zhang J, Hu Y (2020) Sealing performance and mechanical behavior of PEMFCs sealing system based on thermodynamic coupling. Int J Hydrogen Energy 45(43):23480–23489
5. Zhang J, Xie JX (2020) Effect of reservoir's permeability and porosity on the performance of cellular development model for enhanced geothermal system. Renew Energy 148:824–838. J Pipeline Syst Eng Pract 8(2):04016019. https://doi.org/10.1061/(ASCE)PS.1949-1204.0000259
6. Zhang H, Zhang J, Liu S (2016) Mechanical properties of the buried pipeline under impact load caused by adjacent heavy tamping construction. J Fail Anal Prev 16(4):647–654. https://doi.org/10.1007/s11668-016-0128-8
7. Wang H, Chi Q, Li H, Du W (2014) Development and application status of submarine pipeline materials for oil and gas transportation. Welded Pipe Tube 37(8):25–29. https://doi.org/10.3969/j.issn.1001-3938.2014.08.007
8. Helwany S (2007) Applied soil mechanics: with ABAQUS applications. Wiley, Hoboken, NJ

Chapter 4
Thermal Buckling Characteristic Analysis of Submarine Pipelines

4.1 Introduction

In accordance with the specifications dictated by the production process, the transportation of oil and gas via subsea pipelines requires operation at elevated temperatures and pressures. These operational conditions induce significant thermal stresses within the pipeline materials, potentially affecting the structural integrity and service life of the pipeline system [1]. Constrained by seabed soil, submarine pipelines cannot freely deform to release internal stresses, particularly due to thermal expansion. As pipeline length increases, these stresses accumulate. Once they exceed the soil's resistance, sudden deformation occurs, leading to vertical or lateral thermal buckling. This buckling results from compressive axial forces surpassing the pipeline's critical threshold, causing instability and potentially compromising the pipeline's structural integrity [2].

Pipeline thermal buckling can be categorized into three main types [3]: vertical buckling caused by axial forces, horizontal buckling in pipelines laid on flat seabed due to a lack of horizontal constraints, and combined vertical/horizontal buckling in pipelines laid on uneven seabed, where initial deformations follow the seabed's topography. Both vertical and horizontal thermal buckling induce significant stress, strain, and plastic deformation, potentially leading to cracks, fatigue, and localized buckling. Vertical thermal buckling, driven by temperature gradients and constrained by the seabed, is the most critical failure mode for submarine pipelines, as the inability to accommodate thermal expansion increases axial compressive forces. Noteworthy cases include a oil–gas mixed pipeline in the Danish sector of the North Sea in 1986, which experienced vertical thermal buckling with a 2.6-m height due to an 82 °C temperature differential, and a 2000 incident in the Erskine field, UK, where a subsea pipeline was shut down due to severe buckling under a 150 °C temperature differential [4]. Effective mitigation strategies are essential to prevent such failures and ensure the structural integrity of submarine pipelines.

H. Zhang et al., *Submarine Pipeline Integrity: Assessment of Failure Modes and Advanced Evaluation Techniques*, Ocean Engineering & Oceanography 28,
https://doi.org/10.1007/978-3-031-92092-9_4

Submarine pipelines, essential for transporting offshore oil and gas, are susceptible to vertical thermal buckling and yielding, which can lead to hazardous leaks, resource loss, and environmental degradation. As subsea energy infrastructure grows, understanding pipeline deformation due to buckling and developing mitigation measures are critical for maintaining pipeline integrity and environmental safety. This research holds substantial practical value for the design, operation, and long-term reliability of submarine pipelines.

4.2 Vertical Thermal Buckling Theory

The analytical solutions for vertical thermal buckling of an ideal flat pipeline on both a rigid seabed and a rigid-plastic seabed are derived, with a detailed discussion on the variation patterns and parameter sensitivities [5].

4.2.1 *Vertical Thermal Buckling of an Ideal Straight Pipe on a Rigid Seabed*

Using the energy method that satisfies the large deformation theory of buckling [6], the relational expression between the buckling length L, buckling height v_m, and temperature difference ΔT for vertical thermal buckling of an ideal flat pipeline on a rigid seabed can be derived.

According to the work-energy principle, the work W done by external forces on the pipeline is equal to the elastic strain energy V_ε accumulated within the pipeline:

$$W_{f_a} + W_q = V_{\varepsilon_1} + V_{\varepsilon_2} \tag{4.1}$$

where W_{fa} is the the work done by the axial frictional resistance f_a of the pipeline, W_q is the work done by the pipeline's self-weight q, $V_{\varepsilon 1}$ and $V_{\varepsilon 2}$ is the elastic strain energy in the segments $0 \leq x \leq L/2$ and $L/2 \leq x \leq L$, respectively.

$$\begin{cases} V_{\varepsilon_1} = \int\limits_0^{L/2} \left\{ \frac{1}{2}\left[EA(\varepsilon_1 - \alpha\Delta T)^2 + EIv_1''^2\right] + qv_1 \right\} dx \\ V_{\varepsilon_2} = \int\limits_{L/2}^{L_0/2} \left\{ \frac{1}{2}\left[EA\left(u_2' - \alpha\Delta T\right)^2\right] - f_a u_2 \right\} dx \end{cases} \tag{4.2}$$

where E is the Young's modulus, A is the cross-sectional area of the pipeline, EI is the bending stiffness, ΔT is the temperature difference, α is the thermal expansion coefficient, ε_1 and ε_2 is the deformation vectors in the pipeline segments $0 \leq x \leq L/2$ and $L/2 \leq x \leq L$, respectively; v_1 and v_2 is the y-axis displacements in the pipeline segments $0 \leq x \leq L/2$ and $L/2 \leq x \leq L$, respectively; u_1 and u_2 is the x-axis displacements in the pipeline segments $0 \le x \le L/2$ and $L/2 \le x \le L$, respectively; Among them, $\varepsilon_1 = u_1' + \frac{1}{2}v_1'^2$, $\varepsilon_2 = u_2' + \frac{1}{2}v_2'^2$.

The internal forces generated in the deformed sections of the pipeline form the total potential energy during thermal buckling. By applying the principle of minimizing total potential energy, where the first variation equals zero, the maximum buckling height v_m at the point of buckling can be derived as follows:

$$\delta\{V_{\varepsilon_1} + V_{\varepsilon_2}\} = 0 \tag{4.3}$$

Substituting Eq. 4.2 into Eq. 4.3 and rearranging using the variation formula, we obtain:

$$\begin{cases} (EIv_1'')'' - [EI(\varepsilon_1 - \alpha\Delta T)v_1']' = -q, & 0 \le x \le L/2 \\ EA(\varepsilon_1 - \alpha\Delta T)' = 0, & 0 \le x \le L/2 \\ EA(u_2' - \alpha\Delta T)' = -f_a, & L/2 \le x \le L_0/2 \end{cases} \tag{4.4}$$

From the second equation in Eq. 4.4, the pressure acting on the buckling segment is a constant value P:

$$EA(\varepsilon_1 - \alpha\Delta T) = -P \tag{4.5}$$

The boundary conditions of the pipeline at $x = 0$ and $x = L/2$ are:

$$\begin{cases} v_1'(0) = 0 \quad v_1'''(0) = 0 \\ v_1(L/2) = 0, \quad v_1'(L/2) = 0, \quad v_1''(L/2) = 0 \\ v_2(L/2) = 0, \quad v_2'(L/2) = 0, \quad v_2''(L/2) = 0 \\ u_1(L/2) = u_2(L/2), \quad u_1'(L/2) = u_2'(L/2) \end{cases} \tag{4.6}$$

Let $n^2 = P/EI$, and substituting the boundary conditions Eq. 4.6 into Eq. 4.4, we get:

$$v_1(x) = \frac{q}{EIn^4}\left(1 + \frac{n^2L^2}{8} - \frac{n^2x^2}{2} - \frac{\cos nx}{\sin nL/2}\right) \tag{4.7}$$

Substituting the boundary condition at $x = L/2x$ (Eq. 4.6) into Eq. 4.7, we obtain:

$$\tan nL = nL \tag{4.8}$$

Since the maximum buckling deformation occurs at $x = 0$, substituting $x - 0$ and Eq. 4.8 into Eq. 4.7, the maximum buckling height v_m during vertical thermal buckling of the pipeline is:

$$v_m = 2.41 \times 10^{-3}\frac{qL^4}{EI} \tag{4.9}$$

Substituting Eq. 4.8 into $n^2 = P/EI$:

$$P = 20.19\frac{EI}{L^2} \tag{4.10}$$

When $v_2(x) = 0$, $\varepsilon_2 = u_2'$ substituting it into the third of Eq. 4.4:

$$EAu_2'' = -f_a,\quad L/2 \le x \le L_0/2 \tag{4.11}$$

From Eq. 4.5:

$$EA\left(u_1' + \frac{1}{2}v_1'^2 - \alpha\Delta T\right) = -P,\quad 0 \le x \le L/2 \tag{4.12}$$

$$u_1(x) = \left(\alpha\Delta T - \frac{P}{EA}\right)x - \frac{1}{2}\int_0^x v_1'^2(x)dx \tag{4.13}$$

From the boundary condition at $x = L/2$ Eqs. 4.6 and 4.12, we solve Eq. 4.11:

$$u_2(x) = \left[\frac{f_a}{2EA}\left(x + L_a - \frac{L}{2}\right) - \left(\alpha\Delta T - \frac{P}{EA}\right)\right]\left(\frac{L_0}{2} - x\right) \tag{4.14}$$

Due to the boundary condition $u_1(L/2) = u_2(L/2)$, from Eqs. 4.12–4.13, we get:

$$\left(\alpha\Delta T - \frac{P}{EA}\right)\frac{L}{2} - \frac{1}{2}\int_0^{L/2} v_1'^2(x)dx = \left[\frac{f_a}{2EA}L_a - \left(\alpha\Delta T - \frac{P}{EA}\right)\right]L_a \tag{4.15}$$

Substituting the boundary condition $u_2'(L_0/2) = 0$ into Eq. 4.14, it can be obtained:

$$EA\alpha\Delta T - P = f_a\left(\frac{L_0}{2} - \frac{L}{2}\right) = f_aL_a \tag{4.16}$$

Substituting Eq. 4.16 into Eq. 4.15:

$$(EA\alpha\Delta T - P)^2 + f_aL(EA\alpha\Delta T - P) - f_aEA\int_0^x v_1'^2(x)dx = 0 \tag{4.17}$$

Solving Eq. 4.17, we get the following relationship:

$$EA\alpha\Delta T - P = \frac{f_aL}{2}\left(\sqrt{1 + \frac{6.39 \times 10^{-5}Aq^2L^5}{f_aEI^2}} - 1\right) \tag{4.18}$$

Substituting Eqs. 4.9–4.10 into Eq. 4.18, the relationship between the temperature difference ΔT and the buckling height v_m is obtained:

$$\Delta T = \frac{3.96}{A\alpha}\sqrt{\frac{qI}{v_m E}} - \frac{2.26 f_a}{A\alpha}\sqrt[4]{\frac{v_m I}{qE^3}}\left(1 - \sqrt{1 + 0.12\frac{A}{f_a}\sqrt[4]{\frac{Eq^3 v_m^5}{I^3}}}\right) \tag{4.19}$$

4.2.2 Vertical Thermal Buckling of an Ideal Straight Pipe on a Rigid-Plastic Seabed

When the seabed soil has high strength, as shown in Fig. 4.1a, the analytical solution for vertical thermal buckling on a rigid seabed is applicable, as the seabed provides sufficient restraint to prevent significant vertical deformation. However, in real-world applications, seabed soil strength can be as low as a few kilopascals, which makes the rigid seabed assumption unsuitable, resulting in significant design errors. For an elastic seabed, illustrated in Fig. 4.1b, the pipeline separates from the seabed after a certain displacement, causing the spring support to disengage. This results in a nonlinear governing equation, making it difficult to obtain an analytical solution. To address this complexity, we assume a rigid-plastic seabed model, as depicted in Figs. 4.1c and d, where the seabed maintains a constant soil resistance after initial deformation, independent of further displacement [7].

Figure 4.2 illustrates the deformation and force analysis model for vertical thermal buckling of an ideal straight pipeline on a rigid-plastic seabed. When buckling occurs, segments BC and B′C′ become embedded in the seabed. The soil resistance r_i in the embedded sections must exceed the pipeline's weight q to provide adequate support. As the pipeline moves upward, the soil resistance r_u remains constant along the pipeline's length, unaffected by vertical deformation. For pipelines laid bare on the seabed, $r_u = 0$. In the non-buckling regions, seabed resistance precisely counteracts the pipeline's self-weight. The axial force in the buckling section is denoted by p, while p_0 represents the axial force in regions unaffected by buckling. The axial friction force f_a is constant along the pipeline, and the length of the axially deformed section is given by: $L_a = (p_0 - p)/f_a$.

The governing equations for the buckling section of the pipeline are:

$$EI\frac{d^4 v_1}{dx^4} + p\frac{d^2 v_1}{dx^2} = -(r_u + q)(0 \le x \le L_1) \tag{4.20}$$

$$EI\frac{d^4 v_2}{dx^4} + p\frac{d^2 v_2}{dx^2} = r_i - q(0 \le x \le L_1) \tag{4.21}$$

where p is the axial force in the buckling section of the pipeline, v_1 and v_2 is the vertical displacement of the overhanging section (segment OB) and embedded section

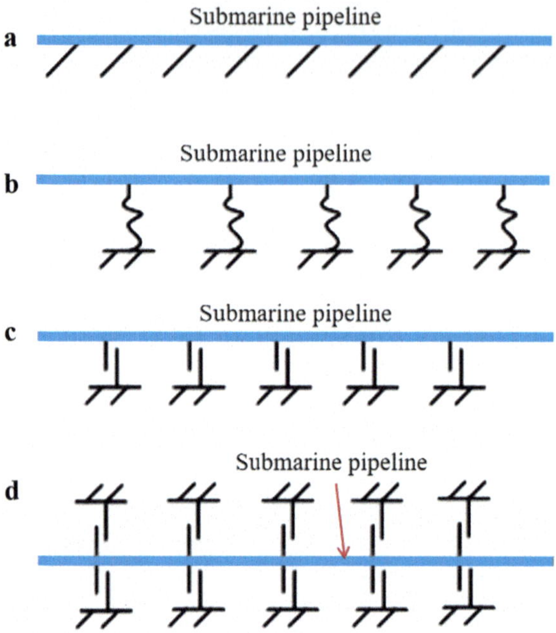

Fig. 4.1 Subsea model **a** On rigid seabed, **b** On elastic seabed, **c** On rigid-plastic seabed, **d** On rigid-plastic seabed

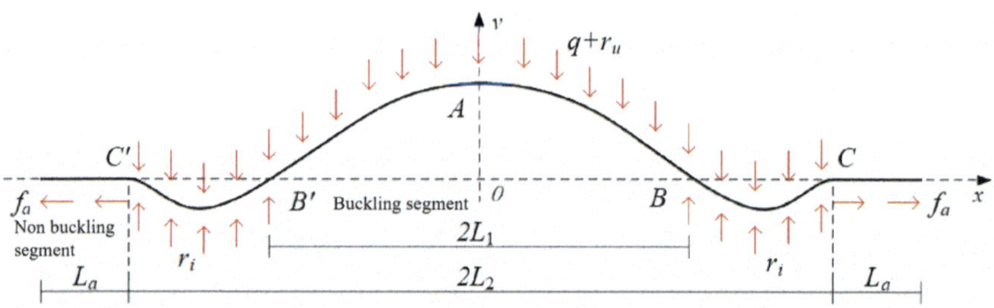

Fig. 4.2 Vertical thermal buckling deformation and stress analysis of an ideal straight pipeline on a rigid plastic seabed

(segment BC), respectively, x is the distance from point O, r_i and r_u is the vertical soil resistance from the seabed during downward and upward pipeline movement, respectively.

The general solutions of Eqs. 4.20–4.21 are:

$$v_1(x) = A_1\cos(nx) + A_2\sin(nx) + A_3x + A_4 - r_0x^2/(2n^2) \tag{4.22}$$

$$v_2(x) = A_5\cos(nx) + A_6\sin(nx) + A_7x + A_8 + Kx^2/(2n^2) \tag{4.23}$$

where $K = \frac{r_i - q}{r_u + q}$, $r_0 = \frac{q}{EI}$, $n = \sqrt{\frac{q}{EI}}$.

Based on the symmetry principle, the slope and shear force at $x = 0$ are zero, and the boundary conditions are:

$$\begin{cases} v_1'(0) = 0 \\ v_1'''(0) = 0 \end{cases} \tag{4.24}$$

Using the continuity conditions for displacement, slope, bending moment, and shear force at $x = L_1$ and $x = L_2$, we obtain:

$$\begin{cases} v_1(L_1) = v_2(L_1), \;\; v_1'(L_1) = v_2'(L_1) \\ v_1''(L_1) = v_2''(L_1), \;\; v_1'''(L_1) = v_2'''(L_1) \\ v_2(L_1) = 0, \;\; v_2'(L_2) = 0 \end{cases} \tag{4.25}$$

From the boundary conditions Eqs. 4.24–4.25, coefficients $A_1 \sim A_8$ can be solved:

$$A_1 = \frac{r_0 \begin{bmatrix} \cos(nL_2)\sin(nL_1) - KnL_2 - nL_1 - KnL_1 + K\cos(nL_2)\sin(nL_1) \\ -\sin(nL_2)\cos(nL_1) - K\sin(nL_2)\cos(nL_1) \end{bmatrix}}{\left[n^4 \sin(nL_2)\right]} \tag{4.26}$$

$$A_2 = 0 \tag{4.27}$$

$$A_3 = 0 \tag{4.28}$$

$$A_4 = \frac{-0.5 r_0 \begin{bmatrix} 2\sin(nL_1) - 2\sin(nL_2) - 2nL_1\cos(nL_2) + 2K\sin(nL_1) \\ -2K\sin(nL_2) - 2KnL_1 nL_2 \sin(nL_2) + 2nL_1\sin(nL_2) \\ +K(nL_1)^2\sin(nL_2) - 2KnL_1\cos(nL_2) + K(nL_2)^2\sin(nL_2) \\ 2nL_1 nL_2\sin(nL_2) + 2KnL_2\cos(nL_2) \end{bmatrix}}{\left[n^4 \sin(nL_2)\right]} \tag{4.29}$$

$$A_5 = \frac{r_0[\cos(nL_2)\sin(nL_1) + KnL_2 - nL_1 - KnL_1 + K\cos(nL_2)\sin(nL_1)]}{\left[n^4 \sin(nL_2)\right]} \tag{4.30}$$

$$A_6 = r_0(1 + K)\sin(nL_1)/n^4 \tag{4.31}$$

$$A_7 = -r_0(1 + K)L_1/n^2 \tag{4.32}$$

$$A_8 = \frac{-0.5r_0\left[\begin{array}{c}-2KnL_1\cos(nL_2) + K(nL_2)^2\sin(nL_2) - 2nL_1nL_2\sin(nL_2)\\ -2KnL_1nL_2\sin(nL_1) + 2\sin(nL_1) + 2K\sin(nL_1)\\ +2KnL_2\cos(nL_2) - 2nL_1\cos(nL_2)\end{array}\right]}{\left[n^4\sin(nL_2)\right]} \tag{4.33}$$

Based on the known boundary conditions of two cross-sections:

$$\begin{cases} v_2(L_1) = 0 \\ v_2''(L_2) = 0 \end{cases} \tag{4.34}$$

It can be obtained:

$$A_5\cos(nL_1) + A_6\sin(nL_1) + A_7L_1 + A_8 + \frac{Kr_0L_2^4(nL_1)^2}{2(nL_2)^4} = 0 \tag{4.35}$$

$$A_5\cos(nL_2) + A_6\sin(nL_2) - \frac{Kr_0L_2^4}{(nL_2)^4} = 0 \tag{4.36}$$

The axial force released in the buckling section equals the axial stiffness of the pipeline multiplied by the axial strain of the entire buckling section, leading to the following expression:

$$\frac{\int_{-L_2}^{L_2}(p_0 - p)dx + 2\int_0^{L_a}(p_0 - p + f_ax)dx}{2L_2 + 2L_a} = \frac{AE}{2L_2 + 2L_a}\left[\int_{-L_1}^{L_1}\frac{1}{2}(v_1')^2dx + 2\int_{L_1}^{L_1}\frac{1}{2}(v_2')^2dx\right] \tag{4.37}$$

Substituting $p_0 = \alpha EA\Delta \boldsymbol{T}$ and Eqs. 4.26–4.27 into Eq. (5.28) yields:

$$EA\alpha\Delta T = P - L_2f_a + L_2f_a\sqrt{1 + \frac{EA}{L_2^2f_a}\left[\int_0^{L_1}(v_1')^2dx + \int_{L_1}^{L_1}(v_2')^2dx\right]} \tag{4.38}$$

Comparing the effects of friction coefficients and cross-sectional areas on buckling displacement and length reveals that, for the same temperature difference, both displacement and length decrease with increasing friction coefficient, increase with cross-sectional area, and decrease with higher buckling modes.

Comparing the impact of soil strength and buckling direction on vertical thermal buckling shows that higher soil strength makes the rigid-plastic solution more similar to the rigid seabed solution. Upward and downward vertical buckling exhibit similar behavior but occur in opposite directions.

4.3 Numerical Simulation Model

4.3.1 Material and Geometric Properties

The submarine pipeline material is API 5L X65, with material and operational parameters as shown in Table 4.1.

The seabed soil has dimensions of 200 m in length, 8 m in width, and 2 m in depth. The material of the seabed soil is silty clay [7], and the stress–strain relationship of the seabed soil is defined using the Mohr–Coulomb model, with parameters as shown in Table 4.2.

4.3.2 Numerical Model and Validation

This study categorizes the seabed's initial imperfections into four types: one-sided elevation, central elevation, cracked, and pitted seabed. These seabed irregularities cause initial deformations in submarine pipelines post-installation. The seabed model directly incorporates these imperfections, excluding the pipeline's external protective coating. The finite element model and cross-sectional diagrams are illustrated in Fig. 4.3: a one-sided elevated seabed, b centrally elevated seabed, c cracked seabed, and d pitted seabed. Grid sensitivity analysis confirms that the model achieves stable buckling displacement and maximum stress values with 18,400 elements.

To validate the theoretical and numerical methods, the results obtained from this study's analytical and finite element simulations are compared with the analytical solutions of [6, 8, 9]. The comparison of vertical thermal buckling deformation, buckling height, and length shows similar trends across methods. Although minor differences exist, the variations fall within an acceptable range. Therefore, the explicit dynamic FE algorithm used for simulating vertical thermal buckling of submarine pipelines is deemed reliable.

4.4 Results and Discussion

This study examines the dynamic deformation characteristics of vertical thermal buckling in high-temperature, high-pressure submarine pipelines. It analyzes the variation of vertical displacement and stress values with temperature differences during buckling on pitted, cracked, one-sided uplifted, and centrally elevated seabed. Additionally, it explores the variation of hoop stress and strain within the primary buckling segments and evaluates the sensitivity of key parameters influencing vertical thermal buckling, including temperature difference, pressure, diameter-to-thickness ratio, pipe-soil friction coefficient, and initial seabed imperfections.

Table 4.1 Materials and operating parameters of the submarine pipeline

Pipe outer diameter	Wall thickness	Density	Young's modulus	Unit weight	Poisson's ratio	Coefficient of thermal expansion	Yield strength	Initial temperature	Working temperature	Working pressure
m	m	kg/m^3	GPa	kN/m^3	–	$°C^{-1}$	MPa	°C	°C	MPa
D	t		E					T_c	T	p
0.2731	0.0206	7850	206	76.9	0.3	0.000011	448	5	95	20

Table 4.2 Material parameters of seabed soils

Bulk density	Density	Young's modulus	Poisson's ratio	Internal friction angle	Cohesive force	Shear strength	Friction coefficient
kN/m^3	kg/m^3	MPa	–	°	kPa	kPa	–
18.86	1923	5	0.2	18.6	18	6	0.5

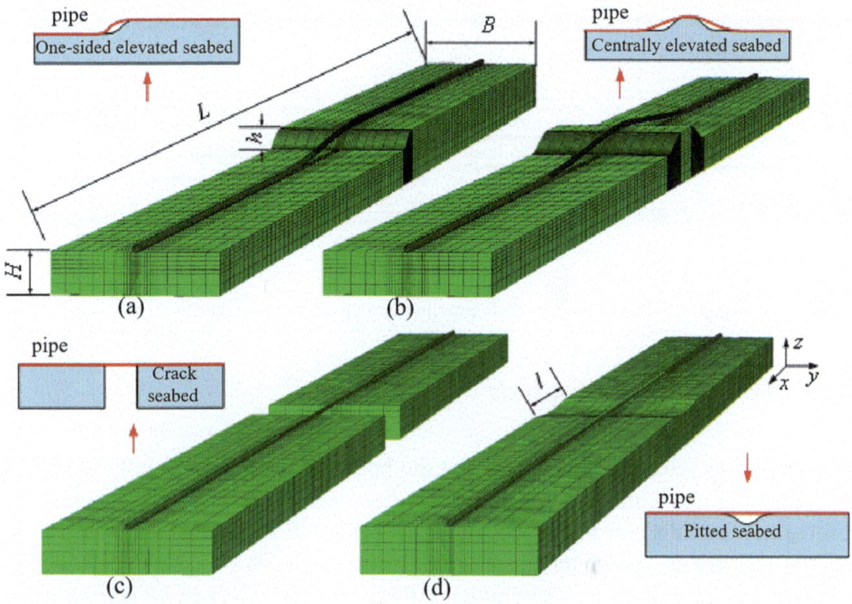

Fig. 4.3 FE model and cross-section diagrams **a** one-sided elevated seabed, **b** centrally elevated seabed, **c** crack seabed, **d** pitted seabed

4.4.1 Deformation Analysis

Figure 4.4 depicts the deformation patterns of buckled pipelines on various seabed types. Different seabed conditions result in distinct buckling configurations in submarine pipelines. The vertical thermal buckling deformations of the pipelines on pitted and cracked seabed are similar to the third-order modal buckling deformation. The pipeline laid on a one-sided uplifted seabed experiences a free span on the one-sided uplifted seabed. For the centrally elevated seabed, the first-order modal buckling deformation occurs. As shown in the figures, buckling deformation intensifies with increasing temperature differences, irrespective of seabed conditions.

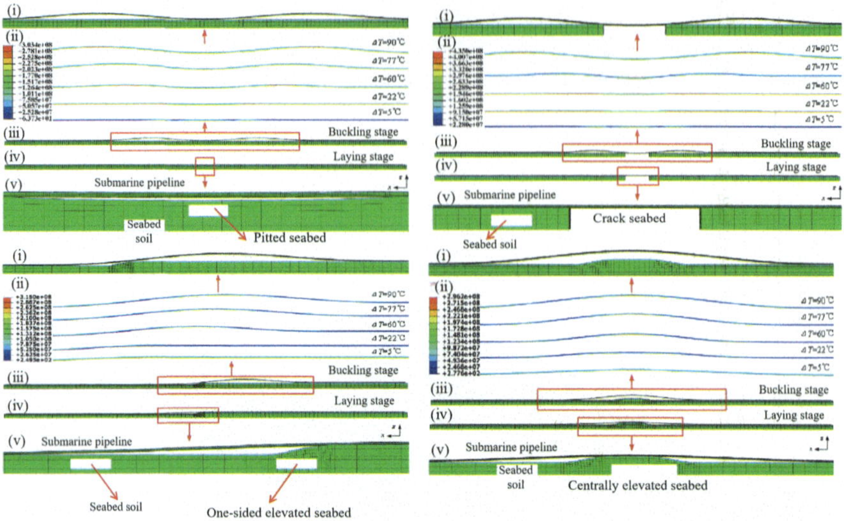

Fig. 4.4 Deformation plots of the buckling pipeline laid on different seabed

4.4.2 Stress–Strain Analysis

As shown in Fig. 4.5, the hoop stress of the buckling pipeline generally increases with the rise in the temperature difference; however, for pipelines laid on one-sided uplifted and centrally elevated seabed, this growth slightly declines as the temperature difference increases significantly.

Pitted seabed: Owing to the symmetric third-order buckling deformation of the submarine pipeline, the left and right segments are identical, allowing analysis to be conducted on the left segment alone. As the temperature difference increases, the hoop stress gradually rises (Fig. 4.5a). The curve transitions from an initial '∞' shape to an arrow-like shape, with the stress variation being most significant in the upper and lower regions of the pipeline. The maximum stress in the lower section rises from 156.18 to 303 MPa, representing a 94% increase.

As the temperature difference increases, the hoop stress in the midsection of the pipeline initially rises and then falls, increasing from 132.1 to 390 MPa and then decreasing to 266.19 MPa. The curve transitions from a circular shape to an "8" shape, with the upper and lower sections showing the most significant stress changes. The maximum stress occurs in the upper section, while the stress in the left and right regions remains nearly unchanged.

Cracked seabed: Fig. 4.5b illustrates the hoop stress distribution in the left segment of the pipeline experiencing vertical thermal buckling on a cracked seabed under varying temperature differences. As the temperature difference increases, the hoop stress rises from 200.13 to 285.83 MPa, an increase of 42.82%. The curve shape changes from an "∞" shape to a heart shape, with stress variations most pronounced in the lower left and lower right sections.

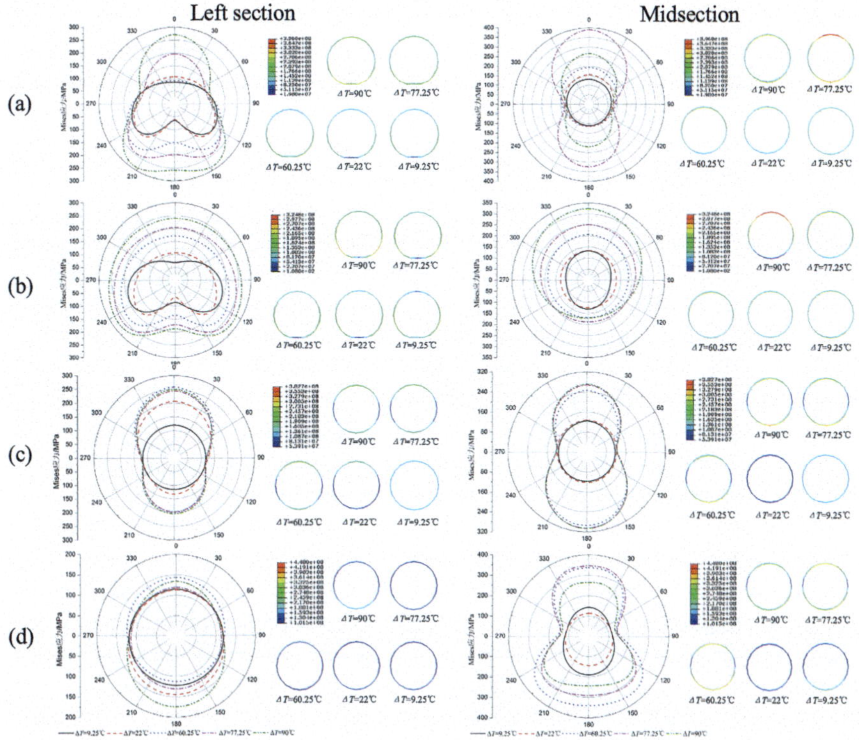

Fig. 4.5 Hoop stress and distribution of the pipeline laid on different seabed **a** pitted seabed, **b** cracked seabed, **c** one-sided elevated seabed, **d** centrally elevated seabed

As the temperature difference increases, hoop stress in the midsection of the pipeline rises from 132.1 to 327.8 MPa, an increase of 148%. The curve transitions from an elliptical to a fan shape, with the upper section exhibiting the highest stress changes and maximum stress, whereas stress variations in the lower section remain minimal.

One-sided elevated seabed: Fig. 4.5c shows the hoop stress in the left segment of the pipeline undergoing vertical thermal buckling on a one-sided elevated seabed at different temperature differences. As the temperature difference increases, the hoop stress gradually rises, but after reaching 60.25 °C, further fluctuations in hoop stress become minimal. The curve shape changes from a circle to an "8" shape, with the upper and lower sections showing the most significant stress changes. The maximum stress occurs in the upper section, while the stress in the left and right regions remains nearly unchanged.

A gradual increase in hoop stress in the midsection of the pipeline can be observed. However, after reaching 77.25 °C, the hoop stress ceases to change further. The curve transitions from a circular to an "8" shape, with the most significant changes in the upper and lower sections, and the maximum stress occurring in the lower section.

Centrally elevated seabed: Fig. 4.5d shows the hoop stress in the left segment of the pipeline undergoing vertical thermal buckling on a centrally elevated seabed. At the upper end of the pipeline, the hoop stress increases with temperature and reaches a maximum of 143.6 MPa at 60.25 °C before decreasing to 134 MPa, a reduction of 6.7%. At the lower end, the hoop stress fluctuates, reaching a minimum of 114 MPa at 60.25 °C and a maximum of 178.5 MPa at 90 °C, an increase of 56.58%.

The hoop stress in the midsection of the pipeline increases initially and then decreases as the temperature difference rises. In the upper section, the stress reaches a maximum at 77.25 °C before decreasing, while in the lower section, the maximum stress occurs at 60.25 °C before decreasing. The curve transitions from an ellipse to an "8" shape, with the upper and lower sections showing the most significant stress changes, and the maximum stress occurring in the lower section. The stress in the left and right sections remains nearly unchanged.

At a temperature difference of 77.25 °C, the maximum hoop stress of 365.6 MPa occurs in the lower left and lower right sections of the buckled pipeline midsection. At 90 °C, the stress decreases to 287.8 MPa, a reduction of 21.28%. A comparison of hoop stress between the midsection and the left segment reveals significantly higher stress in the midsection due to the maximum buckling displacement occurring there.

As shown in Fig. 4.6, the hoop strain of pipelines varies under temperature differences depending on the type of seabed, exhibiting distinct trends and characteristics. Strain distribution and magnitude are influenced by both seabed type and temperature variation, which can be visualized through strain curves and contour plots.

Pitted seabed: Vertical thermal buckling on a pitted seabed increases hoop strain from 0.00047 to 0.00228, a rise of 385.1%. The most significant changes occur in the upper and lower sections, with the upper section experiencing the highest strain. The contour plot highlights higher strain values in the upper, lower-left, and lower-right sections (yellow–red), with lower values in the left and right sections (green–blue).

Cracked seabed: On a crack seabed, hoop strain increases from 0.00104 to 0.0021, representing a 101.9% increase, with the lower-left and lower-right sections showing the largest changes. The contour plot indicates higher strain in the left and right sections (yellow–red) and lower strain in the upper and lower sections (green–blue).

One-sided elevated seabed: Vertical thermal buckling on a one-sided elevated seabed leads to increasing hoop strain, most notably in the upper section. The maximum strain of 0.00211 occurs in the lower section at $\Delta T = 60.25$ °C. The contour plot shows significant strain in the upper and lower sections (yellow-green), with minimal changes in the left and right sections (blue).

Centrally elevated seabed: For centrally elevated seabed, the upper section's hoop strain peaks at 0.002766 at 77.25 °C before decreasing, while the lower section strain continues to rise with increasing temperature. The contour plot shows higher strain in the upper and lower sections (yellow–red), with lower values in the left and right sections (green–blue).

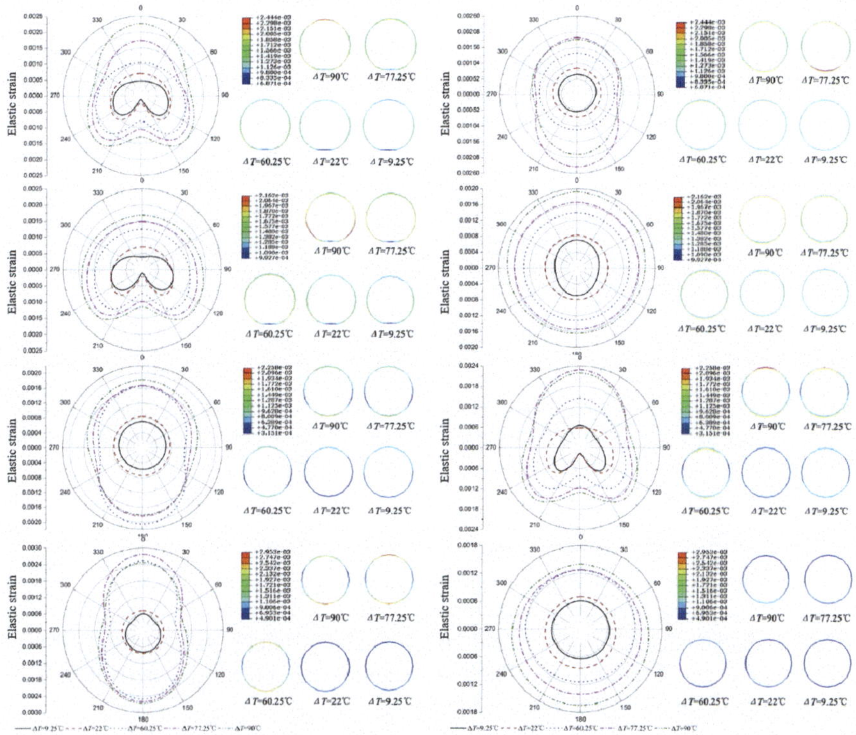

Fig. 4.6 Hoop strain and distribution of the pipeline laid on different seabed **a** pitted seabed, **b** cracked seabed, **c** one-sided elevated seabed, **d** centrally elevated seabed

4.4.3 Sensitivity Analysis

This section discusses the sensitivity analysis of parameters directly affecting the vertical thermal buckling of submarine pipelines. It involves parameters such as temperature difference, pressure, diameter-to-thickness ratio, pipe-soil friction coefficient, dimensions of pockmarked seabed, pipeline burial depth, and their impact on pipeline buckling displacement and pipeline stress values.

4.4.3.1 Temperature Difference

Submarine pipelines experience vertical thermal buckling as temperature differences increase, with deformation and stress intensifying, especially beyond specific temperature thresholds. Figure 4.7 shows variations of subsea pipeline buckling height and axial Von mises stress under different temperature differences and seabed conditions.

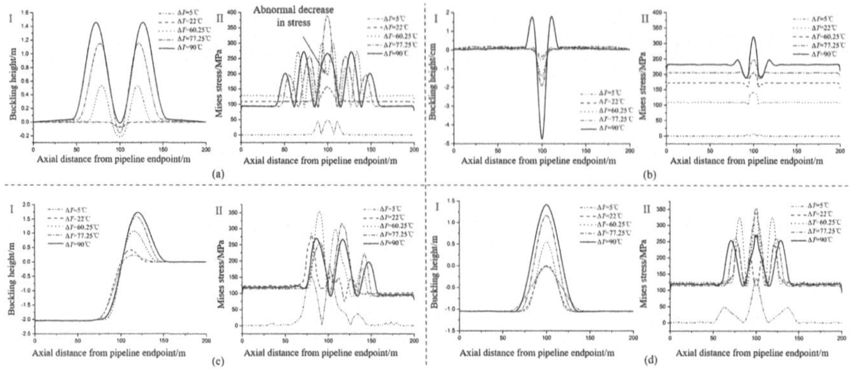

Fig. 4.7 Variations of subsea pipeline buckling height and axial Von mises stress under different temperature differences and seabed conditions **a** pitted seabed, **b** cracked seabed, **c** one-sided elevated seabed, **d** centrally elevated seabed

Pitted seabed: Buckling height and length increase sharply after 22 °C, reaching a maximum height of 1.46 m and a length of 104 m at 90 °C. Stress peaks at 390 MPa at 77.25 °C, concentrated in the buckling section.

Cracked seabed: Deformation and stress rise with temperature, particularly at higher differences. The maximum buckling height is 1.77 m, with a length of 38.4 m, and stress peaks at 322 MPa in the buckling section.

One-sided elevated seabed: Significant buckling begins at 5 °C, reaching a maximum height of 1.733 m and length of 80 m at 90 °C. Stress varies in the buckling section, peaking at 318 MPa.

Centrally elevated seabed: Buckling height and length increase from 5 °C, with a maximum height of 1.416 m and length of 65.2 m at 90 °C. Stress peaks at 354.4 MPa in the buckling section.

4.4.3.2 Internal Pressure

Submarine pipeline deformation and stress vary with increasing pressure across different seabed topographies, with significant changes occurring in the buckling sections, while non-buckling sections remain stable. Figure 4.8 shows variations of subsea pipeline buckling height and axial Von mises stress under different internal pressure and seabed conditions.

Pitted Seabed: As pressure increases, the overhanging and left segments of the pipeline deform upward, with the buckling height rising from 1.19 to 1.53 m and the length increasing from 96 to 112 m. Stress is concentrated in the buckling section, shifting from the center to the sides as pressure increases.

Cracked Seabed: On the cracked seabed, downward buckling displacement intensifies from 0.5 m to 4.745 m as pressure increases. Buckling height and length grow, and stress reaches a maximum of 322 MPa in the buckling section.

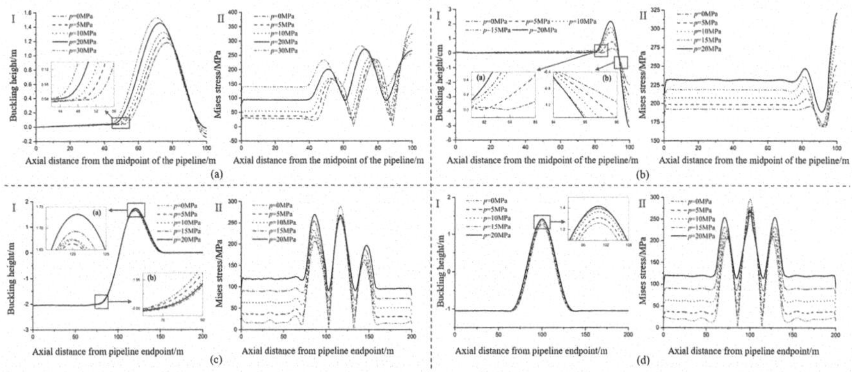

Fig. 4.8 Variations of subsea pipeline buckling height and axial Von mises stress under different pressures and seabed conditions **a** pitted seabed, **b** cracked seabed, **c** one-sided elevated seabed, **d** centrally elevated seabed

One-Sided Elevated Seabed: Upward buckling is minimal, with buckling length increasing from 78 to 80 m and displacement from 1.67 to 1.73 m. Stress increases with pressure but decreases on the right side of the buckling section, dropping from 289 to 260 MPa.

Centrally Elevated Seabed: Upward buckling deformation is slight, with buckling length increasing from 60.8 to 65.2 m and displacement from 1.26 to 1.42 m. Stress initially increases but decreases in the middle of the buckling section, dropping from 296 to 272 MPa.

4.4.3.3 Wall Thickness-Radius Ratio

The buckling deformation and stress of submarine pipelines are affected by the diameter-to-thickness ratio (*DTR*) under various seabed conditions. As *DTR* increases, both buckling height and length tend to increase, particularly in the buckling section, where stress is also higher and increases with *DTR* (Fig. 4.9). Despite differences across seabed types, the overall trend remains consistent.

Pitted seabed: With increasing *DTR*, buckling height increases from 1.39 m to 1.52 m, and length from 108 to 116 m. The pipeline deforms downward in the buckling section and upward at both ends, with the maximum stress rising from 263.6 to 293.8 MPa.

Crack seabed: Buckling height increases slightly with *DTR*, while upward displacement rises from 1.71 to 2.12 m, and downward displacement from 4.707 to 4.874 m. Buckling length remains unchanged, but maximum stress increases from 317 to 376 MPa.

One-sided elevated seabed: As *DTR* increases, stress in the buckling section rises from 206 to 270 MPa, while stress on the right side decreases from 289 to 260 MPa.

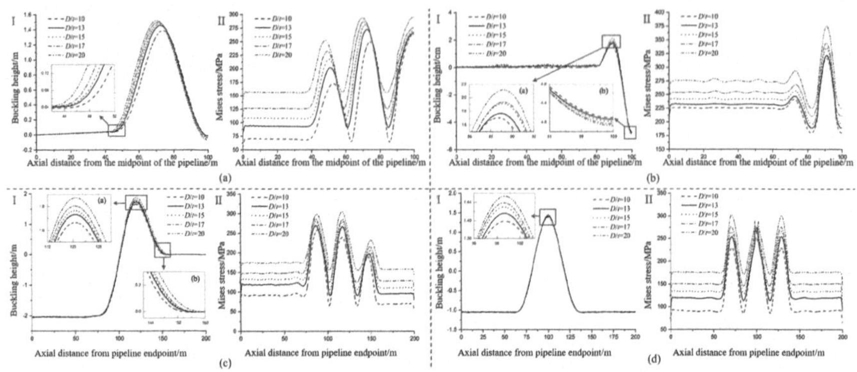

Fig. 4.9 Variations of subsea pipeline buckling height and axial Von mises stress under different *DTR* and seabed conditions **a** pitted seabed, **b** cracked seabed, **c** one-sided elevated seabed, **d** centrally elevated seabed

The maximum stress occurs in the left segment when *DTR* is less than 13, and in the right segment when it exceeds 13.

Centrally elevated seabed: Buckling height and length increase slightly with *DTR*, with upward displacement rising from 1.4 to 1.45 m, and length from 64.8 to 65.2 m. Stress increases, with the left segment rising from 229 to 302 MPa, and the middle segment rising from 252 to 290 MPa.

4.4.4 Seabed Defect Size

The size of seabed irregularities significantly affects the buckling deformation and stress distribution of submarine pipelines. As shown in Fig. 4.10, larger defects typically lead to increased deformation, with stress concentrated in the buckling sections, while non-buckling sections show minimal stress variation.

Pitted seabed: Increasing pitting depth causes the buckling length to expand from 108 to 120 m (+ 11.11%) and the buckling height to decrease from 1.457 m to 1.182 m (− 18.87%). Stress, concentrated in the buckling section, rises from 266.2 to 350 MPa (+ 31.48%).

Crack seabed: As crack length increases, buckling height and length both grow, with more pronounced changes in height. Stress in the buckling section rises sharply, from 321.53 to 670.33 MPa (+ 108.5%).

One-sided elevated seabed: Greater elevation height leads to an increase in buckling height from 1.69 m to 2.72 m (+ 61%). Stress is concentrated in the buckling section, particularly in the right segment, rising from 234 to 265 MPa (+ 13.2%).

Centrally elevated seabed: As elevation length increases, buckling height decreases slightly from 1.44 to 1.32 m (− 8.3%). Stress is concentrated in the middle

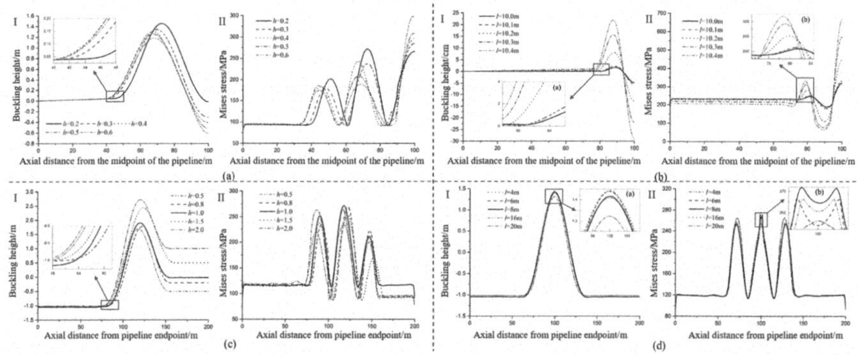

Fig. 4.10 Variations of subsea pipeline buckling height and axial Von mises stress under different seabed defect size and seabed conditions **a** pitted seabed, **b** cracked seabed, **c** one-sided elevated seabed, **d** centrally elevated seabed

of the buckling section, with a complex trend showing an initial increase followed by a decrease.

4.5 Conclusions

This chapter explores the vertical thermal buckling behavior of high-temperature, high-pressure submarine pipelines through theoretical calculations and numerical simulations. Analytical solutions are derived for both rigid and rigid-plastic seabed conditions, and the sensitivity of pipeline parameters is analyzed. By incorporating realistic seabed topography and using finite element methods, the vertical thermal buckling process of submarine pipelines is simulated, and the effects of key parameters are evaluated.

1. For a rigid seabed, an analytical relationship between buckling length L, buckling height v_m, and temperature difference ΔT is derived. Results indicate that buckling displacement and length increase with cross-sectional area and decrease with higher buckling modes, showing similar trends for first- and third-order buckling, as well as second- and fourth-order buckling.
2. Under rigid-plastic seabed conditions, expressions for vertical thermal buckling were derived. As seabed strength increases, the K value in the analytical solution approaches that of a rigid seabed. Upward and downward buckling exhibit similar mechanics but opposite results, with the pipeline embedment ratio decreasing as K increases.
3. Finite element simulations using the ABAQUS explicit dynamic algorithm were conducted to examine the dynamic deformation characteristics of vertical thermal buckling under high-temperature, high-pressure conditions. The simulations demonstrated high reliability and accuracy.

4. In pitted seabed conditions, the pipeline exhibits third-order buckling, with midsection stress significantly higher than in the left section, though strain is similar. Temperature and pipeline pressure affect buckling height and length.
5. In cracked seabed conditions, third-order buckling is observed, with midsection stress higher than in the left section. Buckling deformation increases rapidly when the temperature difference exceeds 22 °C.
6. On one-sided elevated seabed, first-order buckling occurs, with stress and deformation in the right section significantly exceeding those in the left. As elevation height increases, buckling height and length also increase.
7. In centrally elevated seabed conditions, first-order buckling is observed, with midsection stress and strain greater than in the left section. Elevation length has minimal impact on buckling deformation and stress.

This chapter is adapted from the article 'Thermal vertical buckling of surface-laid submarine pipelines on a sunken seabed,' originally published in *Ocean Engineering* (2019), DOI: https://doi.org/10.1016/j.oceaneng.2019.01.005. Reproduced with permission from Elsevier.

References

1. Cai J, Le Grognec P (2022) Lateral buckling of submarine pipelines under high temperature and high pressure—a literature review. Ocean Eng 244:110254. https://doi.org/10.1016/j.oceaneng.2021.110254
2. Liu R, Guo S, Yan S (2015) Study on the lateral soil resistance acting on the buried pipeline. J Coastal Res 73:391–398. https://doi.org/10.2112/SI73-069.1
3. Veritas DN (2007) Global buckling of submarine pipelines, structural design due to high temperature/high pressure. RP-F110, Oslo, Norway
4. Howard JA, Trevisan R, McSpadden A, Glover S (2021) History, evolution, and future of casing design theory and practice. Paper presented at the SPE annual technical conference and exhibition, Dubai, UAE, Sept 2021. https://doi.org/10.2118/206183-MS
5. Hobbs RE (1981) Pipeline buckling caused by axial loads. J Constr Steel Res 1(2):2–10. https://doi.org/10.1016/0143-974X(81)90027-4
6. Liu R et al (2014) Global lateral buckling analysis of idealized subsea pipelines. J Central S Univ 21.1: 416–427.https://doi.org/10.1007/s11771-014-1955-y
7. Liu R, Yan S (2013) Brief history of upheaval buckling studies for subsea buried pipeline. J Pipeline Syst Eng Pract 4(3):170–183. https://doi.org/10.1061/(ASCE)PS.1949-1204.0000127
8. Hobbs RE, Liang F (1989) Thermal buckling of pipelines close to restraints
9. Taylor N, Gan AB (1986) Submarine pipeline buckling—imperfection studies. Thin-Walled Struct 4(4):295–323. https://doi.org/10.1016/0263-8231(86)90035-2

Chapter 5
Mechanical Properties of Steel Strip Reinforced Composite Pipes Under Complex Loads

5.1 Introduction

Steel strip wound reinforced composite pipe (SSRCP) is a robust pipeline solution that ingeniously blends the advantages of composite materials with the strength of steel, utilizing helically wound steel strips to bolster its rigidity [1–3]. This design not only imparts exceptional strength and durability, enabling the pipeline to withstand substantial pressures and stresses, but also ensures it remains lightweight and resistant to corrosion, which is crucial in harsh environments. The pipeline's special features extend to its remarkable flexibility and longevity, significantly reducing transportation and installation costs, and necessitating fewer replacements or repairs over time. Particularly in submarine applications, SSRCP excels due to its ability to resist the corrosive effects of saltwater, withstand the immense pressures found at great ocean depths, and maintain flexibility for installation across the varied and challenging ocean floor terrain. These attributes collectively make SSRCP a reliable, durable, and economically efficient solution for underwater infrastructure, demonstrating the material's innovative application in advancing submarine pipeline technology.

5.2 Numerical Simulation Model

SSRCP includes a leak-proof and corrosion-resistant inner PE pipe, two layers of spirally opposite steel reinforcement belts that resist internal and external pressure, a protective layer that reduces friction, and an outer PE pipe that provides external corrosion resistance [4]. Using a 1100 mm-long composite pipe as the research subject, a numerical calculation model is established as shown in Fig. 5.1 Solid elements are used to model the PE pipe, while shell elements are used to model the steel pipe, while shell elements are used to simulate steel pipe. Due to the non-linear contact between the layers of SSRCP, radial hard contact and normal penalty

H. Zhang et al., *Submarine Pipeline Integrity: Assessment of Failure Modes and Advanced Evaluation Techniques*, Ocean Engineering & Oceanography 28,
https://doi.org/10.1007/978-3-031-92092-9_5

contact are applied to define the contact interactions between the layers. The friction coefficient between the steel strips is 0.18, and that between the steel strips and the PE pipe is 0.22. One end of the pipe is completely constrained, and an axial tensile displacement is applied to the other end. The geometric parameters and material properties are shown in Table 5.1 [5].

Reference [4] proposed a simplified analytical model for calculating the axial tensile stiffness of SSRCP. It can be seen that the axial tensile force of the steel strip is:

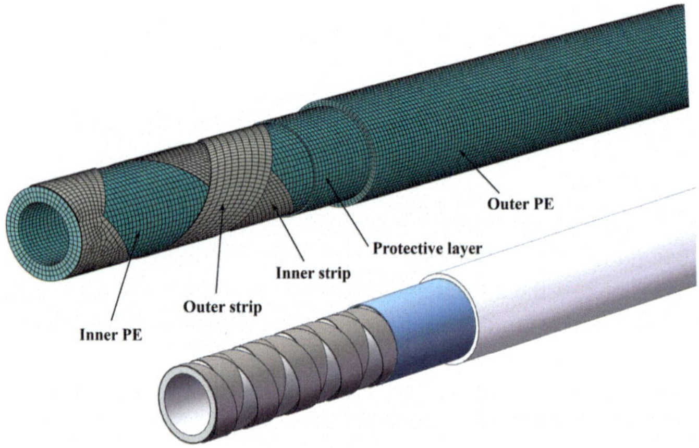

Fig. 5.1 Structure and numerical calculation model of steel tape winding reinforced composite pipe

Table 5.1 Geometric and material parameters

Model	Inside radius (mm)	Wall thickness (mm)	Helical angle (°)	The strip width (mm)	Elastic modulus E (GPa)	Poisson's ratio (μ)	Yield strength (MPa)
Inner PE	25	6	–	–	1.040	0.4	20
Inner steel strip	31	0.5	54.7	52	199	0.26	850
Outer steel strip	31.5	0.5	−54.7	52	199	0.26	850
Protective layer	32	1	–	–	1.040	0.4	20
Outer PE	33	4	–	–	1.040	0.4	20

$$\begin{cases} F_{\text{steel,i}} = nbt\mathrm{E}\cos\theta \cdot \left(\frac{\Delta L}{L}\cos^2\theta + \frac{u_{Ri}}{R_{mi}}\sin^2\theta \right)\{i = 1, 2\} \\ u_{Ri} = -\frac{\Lambda_1\Lambda_5 + \Lambda_6}{1 + \Lambda_2\Lambda_5} \\ \Lambda_1 = \beta E\sin^2\theta \cdot t\frac{\Delta L\cos^2\theta}{L}\left(\frac{1}{R_{m1}} + \frac{1}{R_{m2}}\right) \\ \Lambda_2 = \beta E\sin^4\theta \cdot t\left(\frac{1}{R_{m1}^2} + \frac{1}{R_{m2}^2}\right) \\ \Lambda_5 = \frac{(1+\nu)R_6\left[(1-2\nu)R_6^2 + R_7^2\right]}{\mathrm{E_S}\left(R_6^2 - R_7^2\right)} \\ \Lambda_6 = \frac{2\left(1-\nu^2\right)P_{\text{out}}R_6^2R_7^2}{E_S\left(R_6^2 - R_7^2\right)} + \nu\frac{R_6\Delta L}{L} \end{cases} \tag{5.1}$$

where, u_{Ri} is the radial displacement. β is the reduction factor when considering the gap. E is the elastic modulus of the steel strip. θ is the spiral angle of steel strip winding. L is the pipe length. ΔL is the stretched length. n is the number of steel strips in same layer in the unit length of the longitudinal section. b is the width of steel strip. $R_{\mathrm{m}i}$ is the average winding radius of i-th reinforcement layer of steel strip. t is the thickness of steel strip. R_6 is the outer radius of the inner PE pipe. R_7 is the inner radius of outer PE pipe. E_{S} is the secant modulus of PE material under the current loading step.

The axial tension F_{PEi} of PE pipe is:

$$F_{\text{PEi}} = \frac{\mathrm{E_S}A_{0i}\Delta L}{\Delta L + L} \tag{5.2}$$

where, A_{0i} is the initial area of the PE pipe section.

The total axial tension F_{T} of SSRCP can be regarded as the accumulation of the contribution values of each layer of the pipe:

$$F_T = \Sigma F_{\text{PE}} + \Sigma F_{\text{steel}} \tag{5.3}$$

As shown in Fig. 5.2, a comparison is made between the numerical model, the experimental results from Ref. [5], and the analytical solution from Ref. [4]. The three results exhibit a consistent trend. At the same time, the numerical model is closer to the experimental result than the analytical model. This is because friction between layers and the bending deformation of the spiral steel strip is not considered in the analytical model. Only axial deformation along the strip length is considered. Moreover, since the boundary conditions in the numerical model are rigid and fixed, the experimental fixture inevitably undergoes slight deformation, which will cause errors. Therefore, the numerical model established in this paper is reliable. Stress distribution of each layer of SSRCP under axial tensile load is shown in Fig. 5.3.

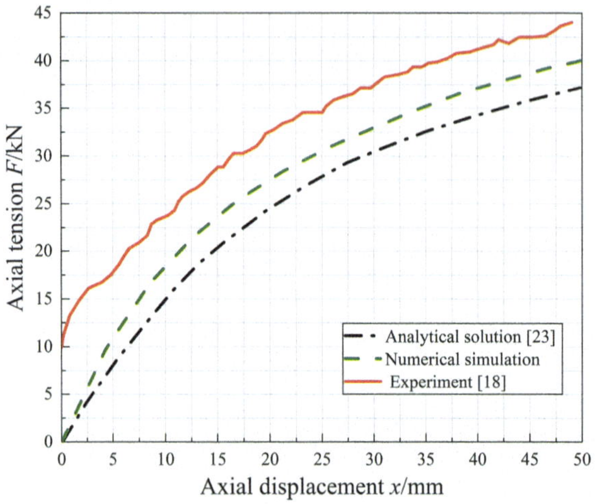

Fig. 5.2 Comparison of the numerical model with the experiment and analytical solution

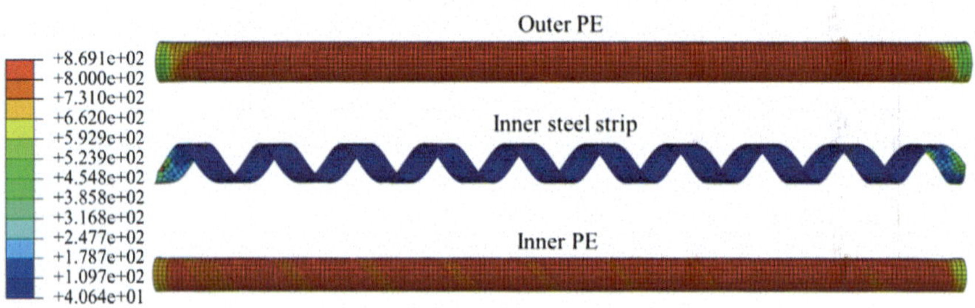

Fig. 5.3 Stress distribution of each layer of SSRCP under axial tensile load

5.3 Analysis of Mechanical Properties

5.3.1 Mechanical Properties of Composite Pipes Under Different Combined Loads

This section analyzes the mechanical behavior of composite pipes under complex loading conditions, including tensile loads, internal and external pressure, and bending loads. The studied load combinations include internal and external pressures with tensile loads, internal and external pressures with bending loads, and pressures combined with both bending and tensile loads, compared to pure tensile and pure bending scenarios. Due to the complex subsea environment, the loading on composite pipes is inherently uncertain [6]. The load-bearing capacity is assessed based on the critical load at which buckling occurs, with higher loads indicating superior performance. The following equation provides the tensile stiffness expression:

$$k = \frac{F}{x} \tag{5.4}$$

where k is the tensile stiffness, F is the axial tensile force, x is the axial displacement.

The bending stiffness can be expressed as:

$$m = \frac{M}{\theta} \tag{5.5}$$

where m is the bending stiffness, M is the bending moment, θ is the bending angle.

Figure 5.4 illustrates the relationship between axial tensile force and elongation for composite pipes. During the elastic phase, tensile stiffness remains constant, while in the transition phase, it decreases with increasing elongation, displaying nonlinear characteristics. Tensile stiffness significantly decreases when external pressure exceeds internal pressure due to the Poisson effect. Bending loads slightly impact tensile stiffness but lower the critical tensile force during buckling. Figure 5.5 shows the bending moment versus bending angle, where bending stiffness under pressurized conditions is significantly lower than under pure bending, owing to variations in interlayer friction. High internal and external pressures improve overall structural integrity but lead to earlier yielding of the outer PE layer, thereby reducing bending stiffness and inducing peak moments at the pipe ends.

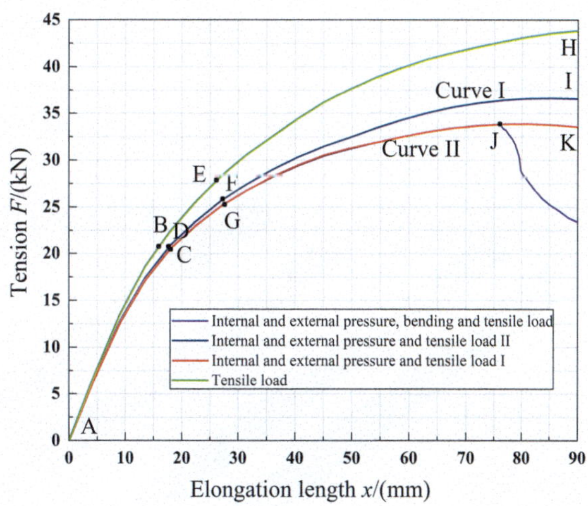

Fig. 5.4 Mechanical properties of composite pipe under different tensile loads

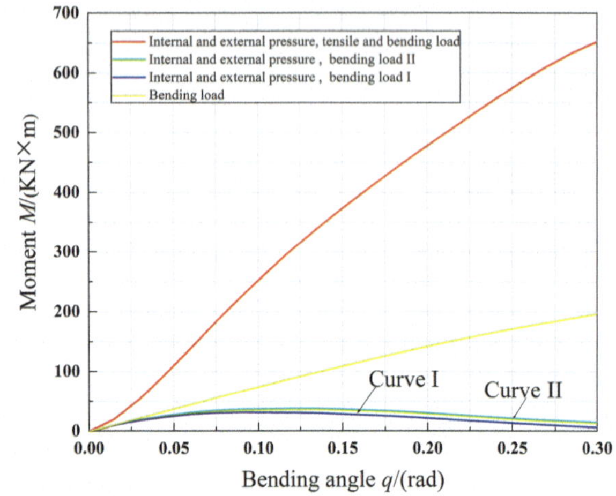

Fig. 5.5 Mechanical properties of composite pipe under different bending loads

5.3.2 Comparison of Mechanical Properties of Each Layer in Composite Pipes Under Complex Loads

Figure 5.6 illustrates the stress distribution in composite pipes under pure tensile loading and combined tensile loading (including internal and external pressure with bending). In pure tension, stress is uniformly distributed, with lower stress values at the ends and a spiral high-stress zone in the inner PE layer, caused by radial compression from the steel strip edges. Under combined tensile loading, severe buckling and failure occur near the free ends of the inner and outer PE layers.

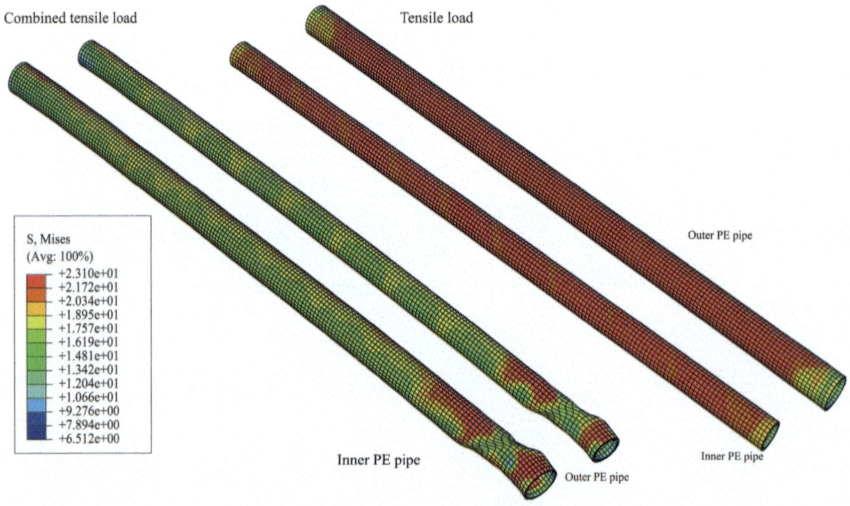

Fig. 5.6 Stress distribution in composite pipes under pure tensile and combined tensile loads

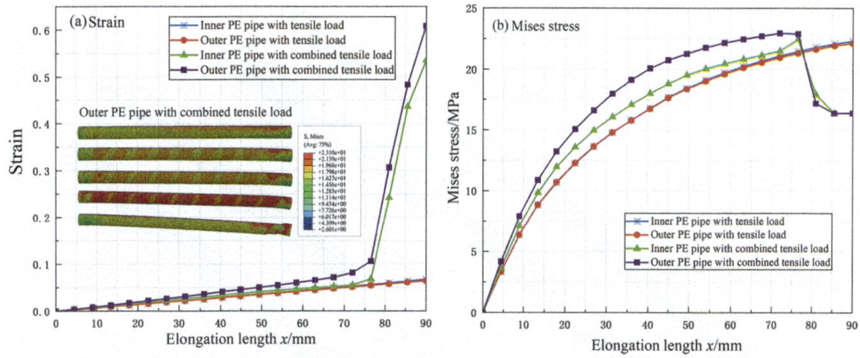

Fig. 5.7 Evolution of stress and strain in the inner and outer PE layers with increasing elongation

Figure 5.7 shows the evolution of stress and strain in the inner and outer PE layers with increasing elongation. Under pure tension, strain remains linear and elastic up to an elongation of 90 mm, with both layers exhibiting consistent strain. Under combined tensile loading, strain increases sharply at 77 mm, reaching the yield point and transitioning into plastic deformation, with the outer layer undergoing greater strain. The Von Mises stress curve indicates that at 76.5 mm elongation, the outer PE layer reaches a maximum stress of 22.60 MPa, while the inner layer reaches 21.51 MPa, highlighting the outer layer's higher sensitivity. A high-stress zone develops at the free end, intensifying in the outer layer's spiral region, ultimately leading to buckling failure at an elongation 77 mm.

Figure 5.8 illustrates the stress distribution in composite pipes under pure bending and combined bending (including internal and external pressures and tensile loads). Under pure bending, high-stress areas are concentrated near the fixed end. In combined bending, stress redistributes toward the free end, leading to buckling failure, with the outer PE layer displaying a more uniform stress distribution.

Figure 5.9 presents the variations in stress and strain with respect to bending angle. Under pure bending, both PE layers remain in the elastic phase. In combined bending, strain sharply increases at 0.24 and 0.25 rad, with the outer layer failing first. In all cases, the outer PE layer consistently exhibits higher Von Mises stress than the inner layer. During deformation, the high-stress zone in the outer layer shifts toward the free end, leading to buckling failure. The increased bending stiffness and decreased flexibility of the composite pipe under combined bending are indicated by the larger curvature radius.

Figure 5.10 presents the stress distribution of the inner wound steel strip under combined tensile and bending loads. The stress distribution pattern remains nearly identical under both loading conditions, exhibiting a non-uniform distribution along the axial direction due to nonlinear boundary effects, which lead to stress concentrations at the strip edges. The stress concentration also follows a symmetric distribution. As shown in Fig. 5.10, A nodal path comprising nodes 1–7 is defined to extract the Von Mises stress values at these locations. According to Ref. [7], under pure tensile

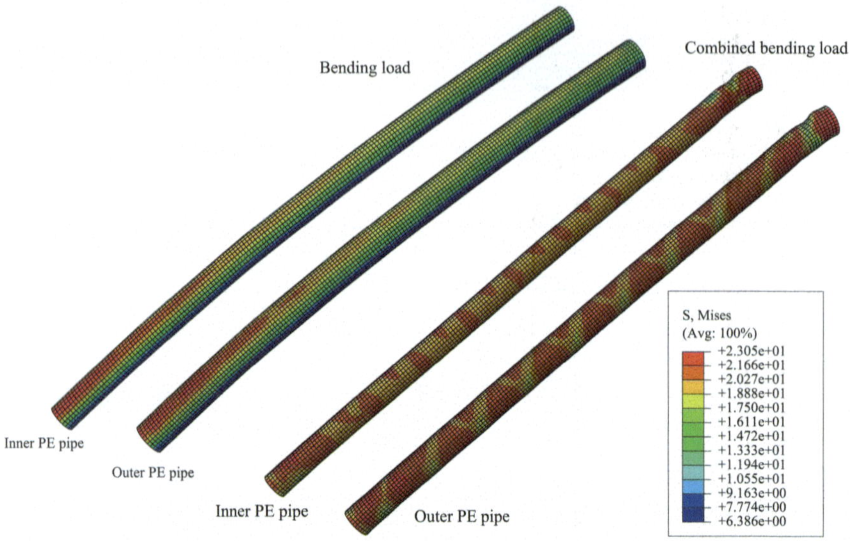

Fig. 5.8 Stress distribution in composite pipes under pure and combined bending

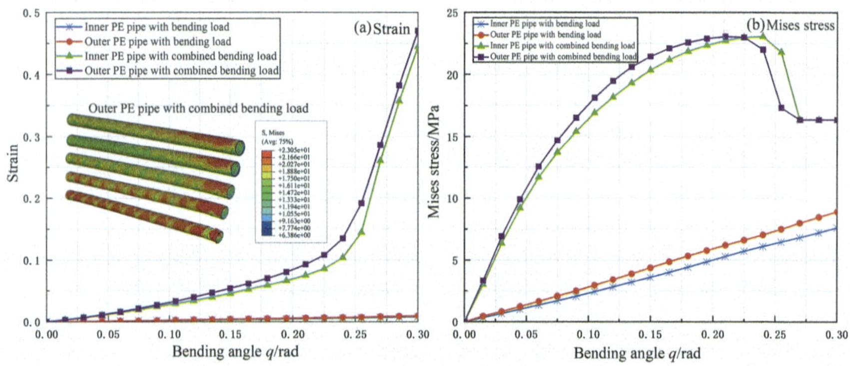

Fig. 5.9 Presents the changes in stress and strain with bending angle

loading, the stress distribution path aligns with the strip width, while under combined loads, the path shifts, indicating that complex loading alters the stress distribution of the inner steel strip.

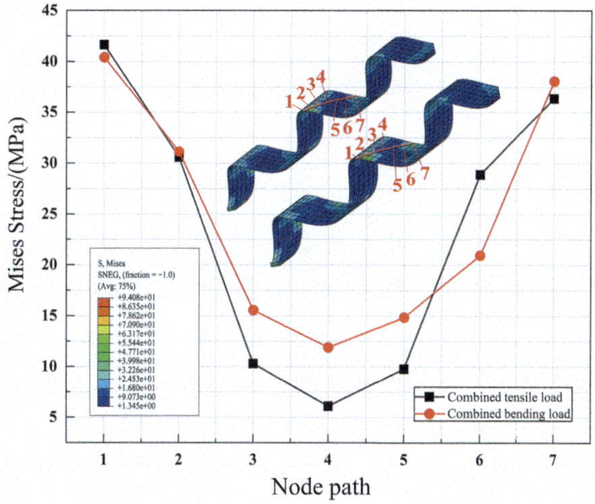

Fig. 5.10 Stress distribution of the inner wound steel strip under combined tensile and bending loads

5.3.3 Sensitivity Analysis

The mechanical performance of composite pipes under complex loads was analyzed by examining variations in the steel strip winding angle and interlayer friction coefficient.

5.3.3.1 Steel Strip Winding Angle

Three winding angles (±54.7°, ± 60.5°, ± 67.7°) were analyzed for steel strip-reinforced composite pipes. As shown in Fig. 5.11, under combined tensile loading, tensile stiffness remains stable during the elastic phase but increases in the transition and yield phases with larger winding angles. This is due to more windings per unit length, which enhances steel strip coverage and friction, reducing interlayer slippage and improving tensile capacity. In combined bending, bending stiffness remains constant in the elastic phase but increases in the yield phase as the winding angle increases, leading to reduced flexibility. The critical bending moment during buckling exhibits a non-monotonic trend, peaking at certain values, underscoring the winding angle's significant impact on composite pipe performance.

5.3.3.2 Coefficient of Friction Between Layers

The analysis of composite pipes with varying friction coefficients (0, 0.1, 0.2, 0.3) indicates that higher interlayer friction significantly impacts their mechanical performance, as shown in Fig. 5.12. Under combined tensile loading, an increased friction

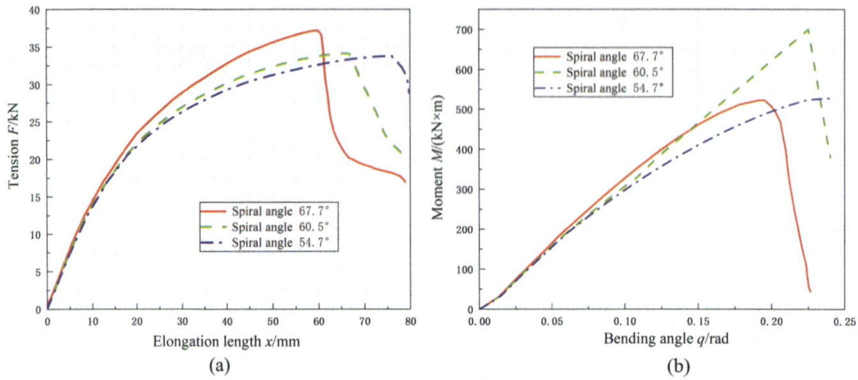

Fig. 5.11 Effect of different helix angles on pipe's mechanical properties under combined loads. **a** Tensile property under combined tensile load, **b** bending property under combined bending load

coefficient enhances tensile stiffness during the transition phase and raises the critical tensile force at buckling, thereby improving the pipe's load-bearing capacity while reducing flexibility. This effect arises because higher friction reduces interlayer slippage, leading to a more uniform distribution of tensile force across the composite layers. Under combined bending loading, a higher friction coefficient similarly increases the pipe's stiffness during the elastic phase, enhancing its resistance to bending deformation. This improvement in bending capacity suggests that the friction between layers plays a critical role in stabilizing the pipe structure under multidirectional stresses. These findings highlight the potential to optimize composite pipe performance by strategically adjusting the interlayer friction coefficient, thus enhancing both tensile and bending load resistance in complex loading environments.

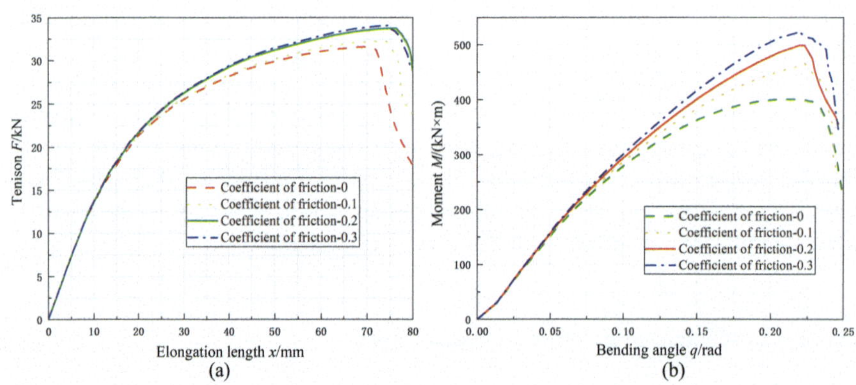

Fig. 5.12 Effect of different friction coefficients on pipe's mechanical properties under combined loads. **a** Tensile property under combined tensile load, **b** bending property under combined bending load

5.4 Nonlinear Hysteretic Behaviors Analysis and Modeling of SSRCP

In this subchapter, the focus is placed on analyzing the nonlinear hysteretic behaviors of SSRCP, a material crucial for shallow sea oil and gas transportation, due to its ability to absorb external energy through deformation, enhancing the stability of marine structures. By considering the nonlinear contact between layers, a finite element model is utilized to delve into SSRCP's behavior under cyclic bending load, with the Bouc-Wen (BW) model being modified through the introduction of an asymmetric factor and intensity enhancement factor for improved accuracy. The model's parameters are meticulously calibrated using stochastic intelligent optimization algorithms, ensuring an accurate representation of the material's hysteretic characteristics. This comprehensive analysis provides vital insights into SSRCP's performance, including its strength and stiffness degradation trends, and the impact of various internal and external factors on its energy absorption capabilities, all of which are integral for optimizing its application in marine environments.

5.4.1 *Hysteresis Analysis*

In this section, we analyze the impact of two loading modes on the hysteresis characteristics of SSRCP, by applying cyclic and reciprocating angular displacements at its free end, as depicted in Fig. 5.13. The relationship between bending moment and angular displacement is illustrated in Fig. 5.14, highlighting the nonlinear hysteretic behavior of SSRCP, attributed to friction, slippage, and complex interlayer interactions. Post-loading, the unloading curve deviates from the loading path, indicating residual bending moments even without deformation, a trend intensifying with subsequent loading cycles.

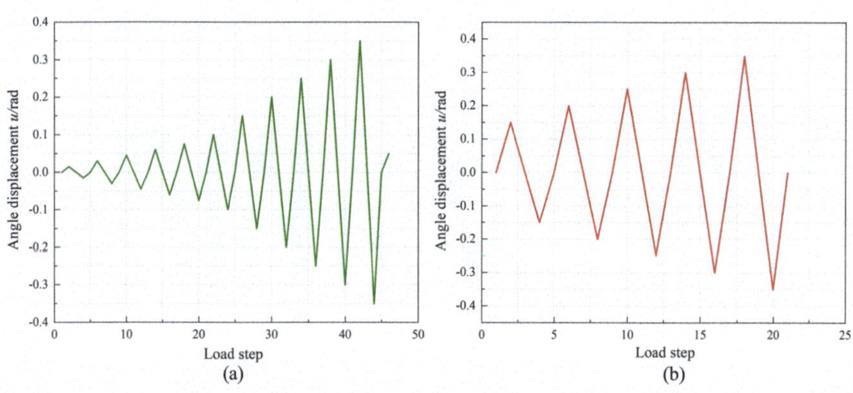

Fig. 5.13 Two loading modes **a** Loading mode I, **b** loading mode II

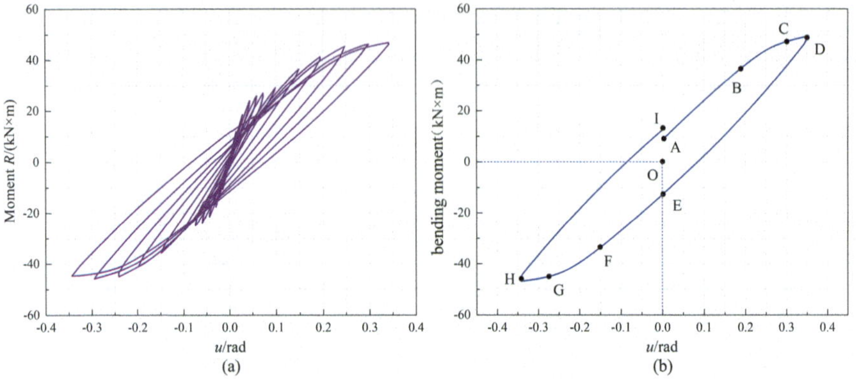

Fig. 5.14 Hysteresis curves of SSRCP at the same displacement amplitude. **a** Complete loading process, **b** last cycle load

The hysteresis curve's gradual rotation towards the horizontal axis signifies a decrease in overall bending stiffness and an increase in SSRCP's flexibility, demonstrating stiffness degradation. Concurrently, the bending moment's growth rate at displacement amplitude decreases, while strength increases. The hysteresis curve's enclosed area represents SSRCP's frictional energy dissipation per cycle, which escalates with more loading cycles and larger bending angles, emphasizing the non-linear relationship between bending moment and displacement.

Figure 5.14b details the hysteresis curve during the last loading cycle, showing distinct loading stages: AB (linear behavior with high stiffness), BC (transition to non-linear with decreasing stiffness), and CD (complete slip with reduced stiffness). The unloading curve (DE) exhibits hysteresis. Upon reverse loading, the pipe transitions through elastic, partial slip, and complete slip stages (EF, FG, GH), leaving a residual bending moment after unloading.

In the final cycle of loading, Fig. 5.15 illustrates the stress distribution across each layer of the pipe when it reaches its limit position. Observations reveal a stress concentration at the pipe's fixed end, with the outer PE pipe being more susceptible to influence from the steel belt during bending.

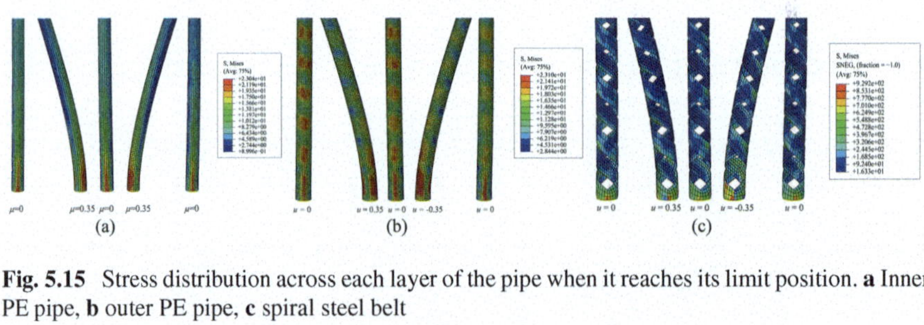

Fig. 5.15 Stress distribution across each layer of the pipe when it reaches its limit position. **a** Inner PE pipe, **b** outer PE pipe, **c** spiral steel belt

Figure 5.16a illustrates the relationship between frictional energy dissipation and bending displacement in SSRCP, showing an increase in dissipation with more loading cycles, and accelerated growth during unloading and reverse loading due to interlayer deformation. Figure 5.16b depicts an exponential growth in frictional energy dissipation at limit displacement, indicating a rapid increase in dissipation rate with more loading cycles and larger limit displacements, slightly more pronounced during reverse loading.

Figure 5.17a shows the hysteresis curve rotating horizontally with more loading cycles, indicating decreased bending stiffness, increased flexibility, and more pronounced non-linear behavior in SSRCP. The enclosed area of the curve, representing energy dissipation, also expands.

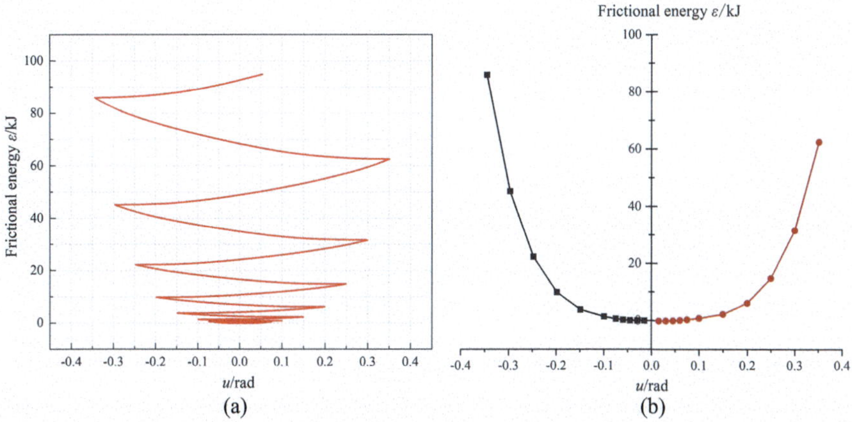

Fig. 5.16 Frictional energy dissipation in SSRCP during cyclic loading. **a** Frictional energy dissipation during loading, **b** energy curve at limit displacement

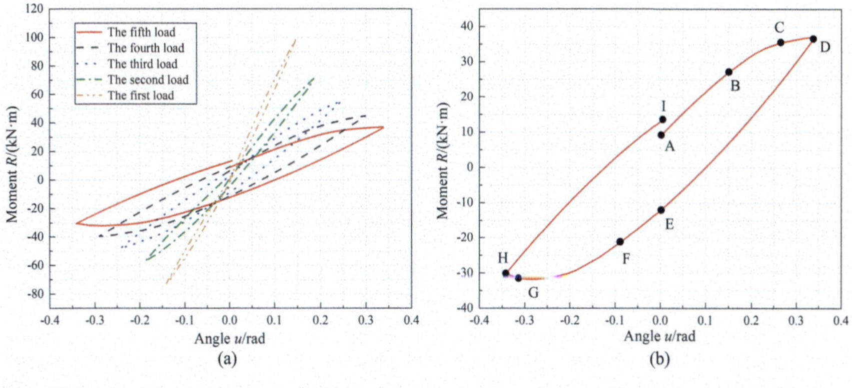

Fig. 5.17 Hysteresis curve of SSRCP at large displacement amplitudes. **a** Complete loading process, **b** last cycle load

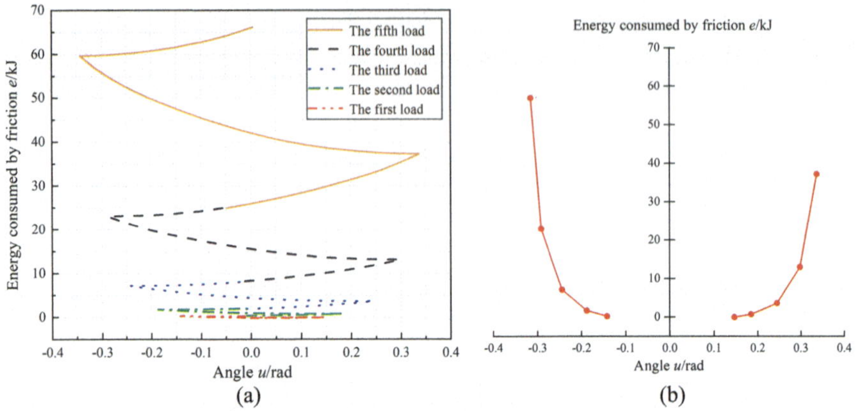

Fig. 5.18 Frictional energy dissipation in SSRCP under bending cycle loading. **a** Frictional energy dissipation during loading, **b** energy curve at limit displacement

In loading mode II, Fig. 5.17b shows the bending moment and angular displacement of SSRCP's free end. During the AB stage, bending stiffness remains high due to significant interlayer friction preventing slippage. In BC, as deformation increases, static friction reaches its limit, leading to partial slippage and decreased stiffness. CD stage sees overall slippage and further reduced stiffness. During the DE unloading stage, hysteresis is observed due to interlayer friction. In the reverse loading phase (EF, FG, GH), the pipe transitions through elastic, partial slip, and complete slip stages, ultimately leaving a residual bending moment.

Figure 5.18a, b show similar trends in frictional energy accumulation and dissipation as Fig. 5.16a, b, with an increase in dissipation with more loading cycles and a rapid growth rate at limit displacement, respectively.

The characteristics in loading mode II are largely similar to those in loading mode I, with the key difference being a decrease in bending moment at displacement amplitude in mode II, indicating strength degradation. This is attributed to the absence of small displacement preloading in mode II, suggesting that SSRCP could benefit from small amplitude preloading post-production to enhance its strength in service.

To further assess the impact of small amplitude preloading on SSRCP's strength, a constant displacement amplitude (0.2 rad) was applied repeatedly, with the bending moment and angle relationship depicted in Fig. 5.19a. Despite the consistent loading, SSRCP exhibited a trend of strength degradation as loading times increased, with the hysteresis curve rotating clockwise around the coordinate origin. This rotation, alongside insights from Fig. 5.17a, indicates a reduction in overall stiffness and showcases stiffness degradation characteristics in SSRCP.

The area enclosed by the hysteresis curve showed minimal change with additional loading, highlighting that under constant conditions, increasing the limit displacement is the primary factor accelerating frictional energy dissipation in SSRCP.

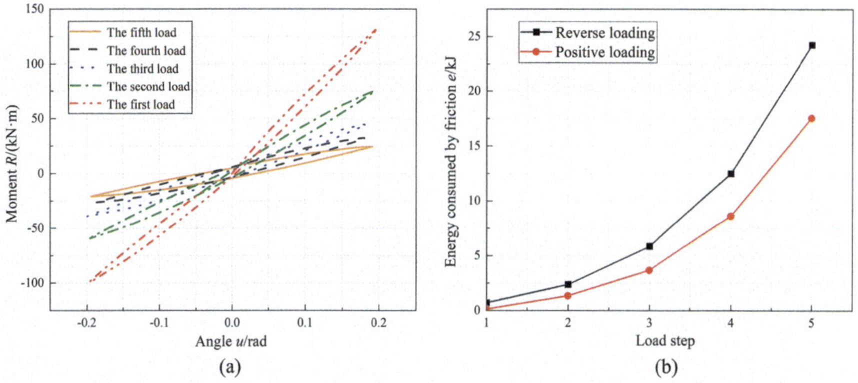

Fig. 5.19 Stiffness degradation and frictional energy dissipation of SSRCP. **a** Stiffness degradation, **b** frictional energy dissipation

Figure 5.19b presents the frictional energy dissipation across loading cycles, revealing a concave curve trend and a higher rate of energy dissipation growth with more loading cycles. Due to early deformation accumulation, the frictional energy dissipation during reverse loading to the limit position consistently exceeded that of forward loading, with this disparity widening as the number of loadings increased.

5.4.2 SSRCP Hysteresis Characteristics Under Different Parameters

The hysteresis behavior of SSRCP originates from interlayer friction during bending deformation. The compression force between these layers significantly influences this friction. In a typical submarine pipeline, the external pressure remains constant. In this study, an external pressure of 3 MPa was applied, while internal pressures of 1, 2, and 3 MPa were tested. Under loading mode I, the hysteresis curve of SSRCP (Fig. 5.20) exhibits significant changes with increasing internal pressure. At internal pressures $\geq$ 2 MPa, no hysteresis loop is observed for small amplitude loading, and the load–displacement curve is nearly linear. As the loading amplitude increases, the hysteresis loop expands; however, bending stiffness remains unchanged, indicating no stiffness degradation. This is attributed to increased internal pressure enhancing friction between pipe layers, thereby improving overall stiffness.

Under loading mode II (Fig. 5.21), higher internal pressures lead to more closely spaced hysteresis curves. A reduced pressure differential mitigates stiffness degradation in SSRCP. With both internal and external pressures at 3 MPa, SSRCP's overall stiffness remains consistent between the fourth and fifth loadings. This is due to increased interlayer pressure, enhancing SSRCP's integrity.

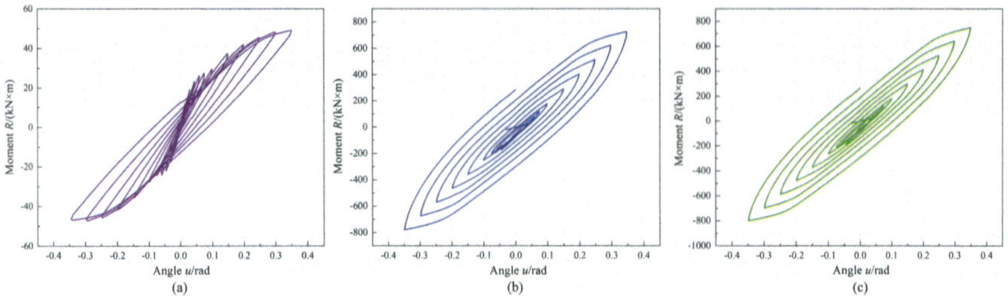

Fig. 5.20 Hysteresis curve of SSRCP under different internal pressures of loading mode I. **a** 1 MPa, **b** 2 MPa, **c** 3 MPa

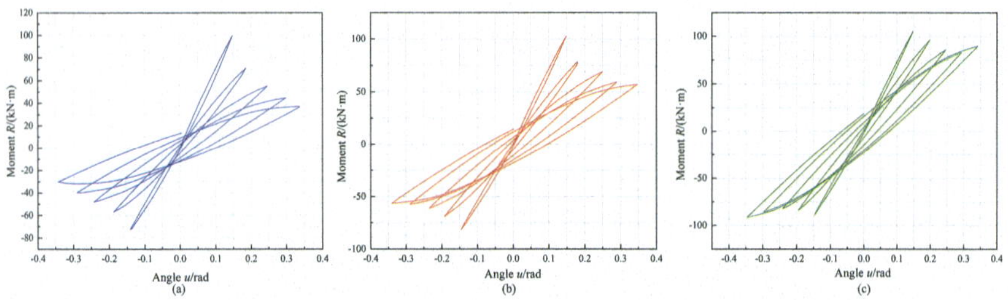

Fig. 5.21 Hysteresis curve of SSRCP under different internal pressures of loading mode II. **a** 1 MPa, **b** 2 MPa, **c** 3 MPa

The skeleton curve (Fig. 5.22) represents SSRCP's stress-deformation relationship under cyclic bending load. It highlights SSRCP's bearing capacity and stiffness. Considering SSRCP's uncertain in-service cyclic load, its strength degradation should be factored in when analyzing hysteresis characteristics. For loading mode II, with internal pressure at 1 MPa and external pressure at 3 MPa, the skeleton curve can be divided into ascending (elastic) and descending (stiffness degradation) parts. The curve's height indicates SSRCP's deformation capacity during low-cycle loading, while its slope indicates the rate of stiffness reduction.

Figure 5.22 also shows that as pressure difference decreases, SSRCP's bending stiffness increases, reducing its flexibility. This effect is more pronounced under reverse loading. In the stiffness degradation stage, SSRCP with 2 MPa internal pressure has significantly lower stiffness than with 1 MPa, especially as ultimate bending displacement expands.

The friction coefficient between the layers is one of the main factors influencing the friction performance. The same loading method as above is adopted to obtain the hysteretic curve with different friction coefficients as shown in Fig. 5.23. With the increasing of friction coefficient, the hysteresis curve rotates counterclockwise as a whole, the stiffness degradation effect is weakened and the overall stiffness of the pipe is increased.

Figure 5.24 displays SSRCP's frictional energy dissipation curve during 0.35 rad to − 0.35 rad loading. Post-cycle loading sees a rapid increase in SSRCP's frictional

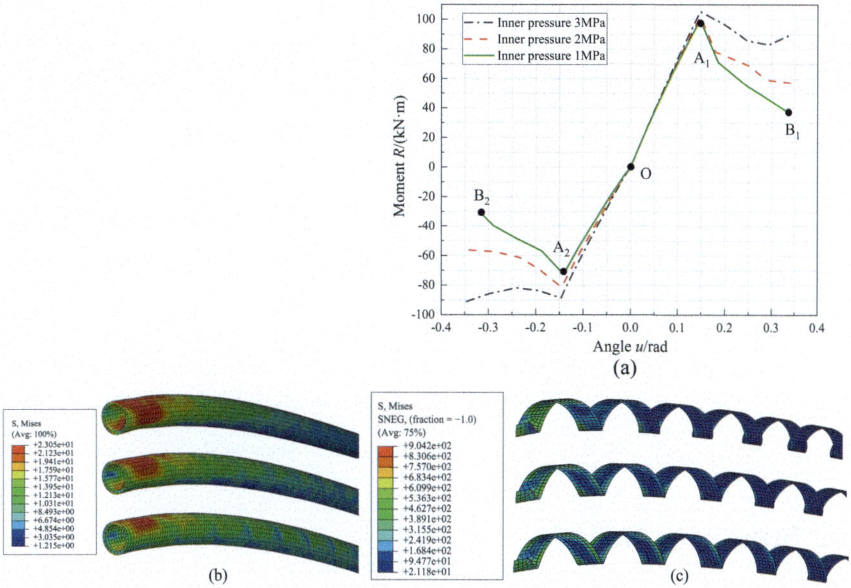

Fig. 5.22 Skeleton curve and stress distribution of SSRCP under different internal pressures

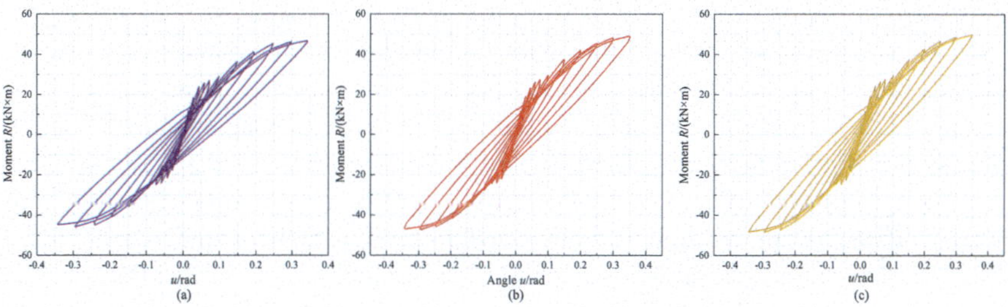

Fig. 5.23 Hysteresis loops of SSRCP under different friction coefficients. **a** 0.18, **b** 0.22, **c** 0.26

energy dissipation, which doesn't linearly correlate with increasing internal pressure. However, the growth trend of frictional energy dissipation remains consistent across different internal pressures.

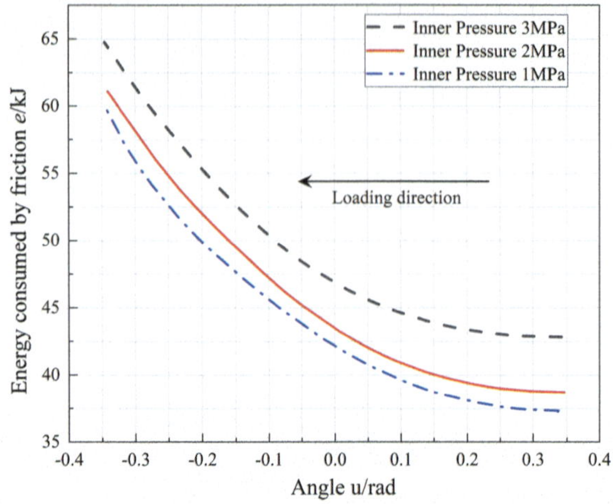

Fig. 5.24 Frictional energy dissipation curve of SSRCP during loading from 0.35 rad to − 0.35 rad

5.5 Sensitivity Analysis

5.5.1 *Diameter-to-Thick Ratio*

The hysteretic behavior of SSRCP is significantly influenced by wall thickness, which is modified by adjusting the inner diameter while maintaining a constant outer diameter (Fig. 5.25). During the elastic phase, a reduction in the diameter-to-thickness ratio increases the slope of the skeleton curve, signifying enhanced SSRCP stiffness and reduced flexibility. In the stiffness degradation stage, the skeleton curve rises with a decreasing diameter-thickness ratio, with reverse loading showing more regularity than forward loading. However, at diameter-to-thickness ratios of 4.11 and 3.70, the corresponding bending moment exhibits minimal increase, and the stiffness degradation rate of SSRCP remains nearly constant under low-cycle loading.

The frictional energy dissipation curve of SSRCP during loading from 0.35 rad to − 0.35 rad is shown in Fig. 5.25b. The frictional energy dissipation of SSRCP increases at an accelerated rate, and as the diameter-to-thickness ratio increases, the frictional energy dissipation of SSRCP decreases. The trend of frictional energy dissipation remains essentially consistent across different diameter-to-thickness ratios.

5.5.1.1 Spiral Winding Angle of the Steel Strip

When the friction coefficient between the steel strip and PE pipe is 0.22, and the helical angles of different steel strips are ± 65.68°, ± 66.91° and ± 68.18° respectively. The spiral winding angle of the steel strip is the included angle between the axial direction and the tangent of the spiral direction as shown in Fig. 5.26. Helix

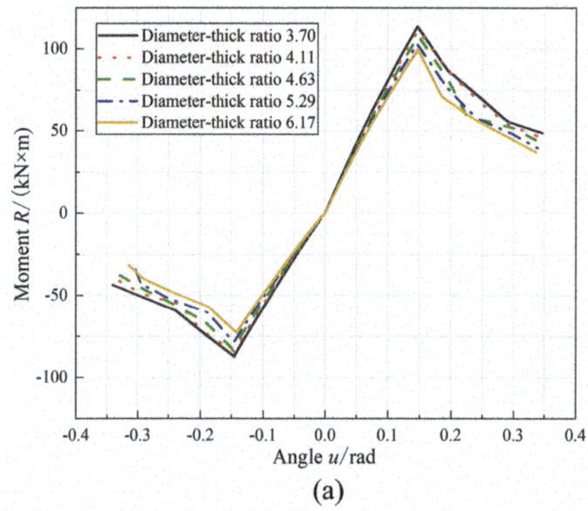

Fig. 5.25 Effect of diameter-to-thickness ratio on SSRCP bending and energy dissipation. **a** Skeleton curves of SSRCP, **b** frictional energy dissipation curve of SSRCP during loading from 0.35 to − 0.35 rad

angle can be expressed as:

$$\theta = \arctan\left(\frac{x_{OM}}{y_{NM}}\right) = \arctan\left(\frac{2\pi R_{mi}}{y_{NM}}\right) \tag{5.6}$$

where, y_{NM} is the gap formed by winding a spiral steel strip.

Figure 5.27a illustrates the skeleton curves of SSRCP with varying steel belt spiral angles, showing that as the spiral angle increases, the elastic stiffness and flexibility of SSRCP also increase during the elastic stage. In the stiffness degradation stage, the skeleton curve rises with the increasing helix angle, indicating enhanced flexibility of SSRCP under low-cycle cyclic loading, though the stiffness degradation rate remains largely unaffected.

Figure 5.27b presents the frictional energy dissipation curve of SSRCP during loading from 0.35 to − 0.35 rad. The frictional energy dissipation is nearly identical for spiral angles of ± 66.91° and ± 68.17° during the unloading phase. However, after reverse loading, SSRCP with a spiral angle of ± 66.91° exhibits slightly lower

Fig. 5.26 Schematic diagram of spiral angle of steel belt

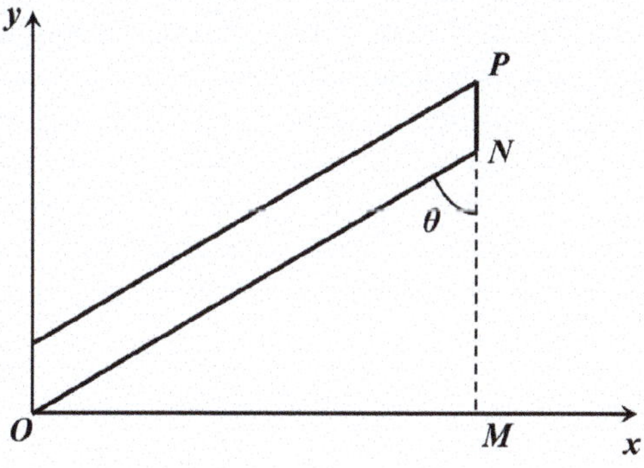

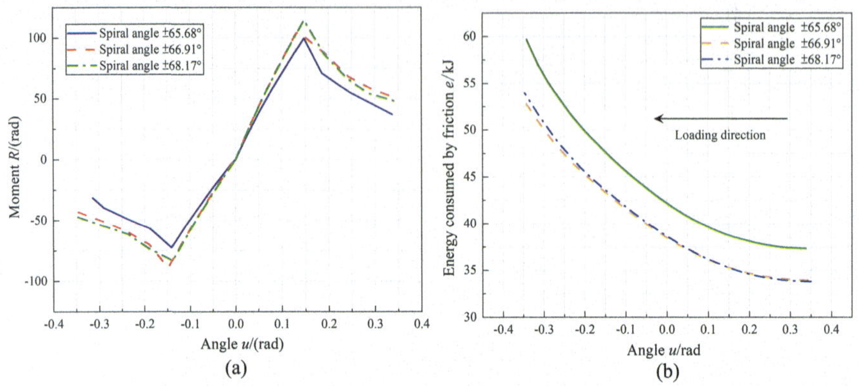

Fig. 5.27 Effect of belt spiral angles on SSRCP bending and energy dissipation. **a** Skeleton curves of SSRCP, **b** frictional energy dissipation curve of SSRCP during loading from 0.35 to − 0.35 rad

frictional energy dissipation than that of ± 68.17°. On the other hand, SSRCP with a helix angle of ± 65.68° shows significantly higher frictional energy dissipation than the other two cases in both unloading and reverse loading, suggesting the existence of a critical helix angle that triggers a substantial change in SSRCP's energy absorption capacity.

5.5.1.2 Friction Coefficient

Figure 5.28a displays SSRCP skeleton curves under different steel strip friction coefficients, showing no impact on stiffness in the elastic phase for both forward and reverse loading. After 0.24 rad in positive loading, higher friction coefficients lead to flatter skeleton curves, indicating a slower stiffness degradation rate during low-cycle cyclic loading.

Figure 5.28b presents skeleton curves with varying friction coefficients between the steel strip and PE pipe, again showing no change in SSRCP stiffness in the elastic stage. After ± 0.23 rad, an increase in friction coefficient results in a decreased stiffness degradation rate.

Figure 5.29 depicts the frictional energy dissipation curves of SSRCP during loading from 0.35 to − 0.35 rad, with increasing interlayer friction coefficients. The frictional energy dissipation of SSRCP rises gradually in low-cycle cyclic loading as friction coefficients increase between steel strips and between steel strips and PE pipe. The trend of increasing frictional energy dissipation aligns with the increase in friction coefficient between steel strips from unloading to reverse loading. However, changes in the friction coefficient between steel strip and PE pipe cause the frictional

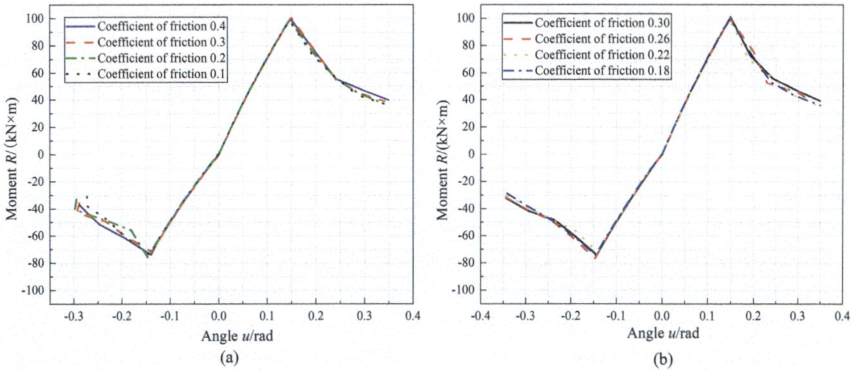

Fig. 5.28 Skeleton curves of SSRCP with different friction coefficients. **a** Steel strips, **b** steel strip and PE pipe

energy dissipation curve to shrink at reverse loading of − 0.35 rad. The friction coefficient between steel strips has a more pronounced effect on SSRCP's energy absorption, suggesting that increasing this friction coefficient can enhance the stability of pipe connections.

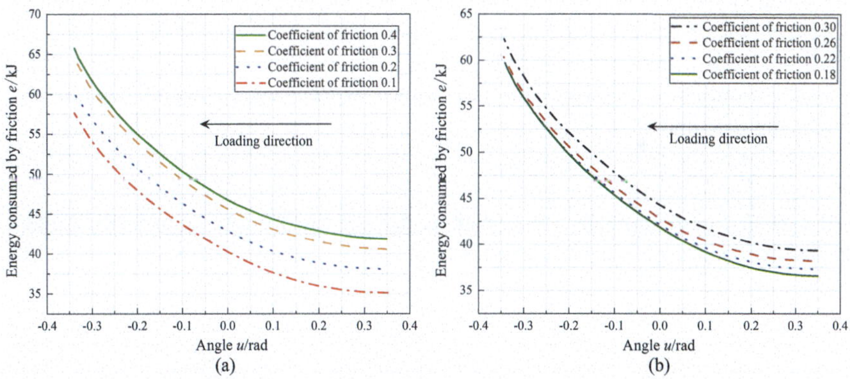

Fig. 5.29 Frictional energy dissipation curve of SSRCP with different friction coefficients. **a** Steel strip and steel strip, **b** steel strip and PE pipe

5.5.2 Nonlinear Hysteretic Modeling and Parameter Identification

5.5.2.1 Establishment of Asymmetric Bouc-Wen Model

Nonlinear hysteretic behavior is common in structures under significant cyclic excitation like earthquakes, winds, or waves [8]. Such systems require "in-memory" models as simple algebraic expressions are insufficient to describe the restoring forces. Pioneering research in this field includes works by [9–15].

The Bouc-Wen model, proposed by Bouc [16] and extended by Wen [17], stands out for modeling structural hysteretic behavior. Baber and Noori [18] further developed this model to incorporate degradation and contraction effects. However, its restoring force–displacement hysteresis curve is symmetric, making it challenging to accurately depict the structure's asymmetric hysteretic characteristics. Addressing this issue, Kwok et al. [19] and Zhu and Wang [20] modified the velocity term and the differential relation between relative lateral displacement and hysteresis displacement, respectively, introducing an asymmetric Bouc-Wen model suitable for hysteresis dampers and piezoelectric ceramics.

The Bouc-Wen model is a smooth hysteretic model, which contains linear stiffness, nonlinear stiffness, and nonlinear damping terms. Therefore, the model can reflect not only the elastic stage and the stage of complete slip but also the transition phase from the elastic stage to the stage of complete slip, which can be used to simulate the interaction between SSRCP layers. Moreover, when the frequency of external excitation changes, the relationship between hysteresis force and displacement does not change, and the friction generated by the interaction between SSRCP layers is independent of the frequency of displacement excitation. Therefore, this model has significant applicability to describe the hysteresis characteristics of SSRCP.

Based on the fundamental Bouc-Wen model, the restoring force of SSRCP under bending load can be described as follows:

$$R(\boldsymbol{u}, z) = (1 - \alpha)k_0 z \tag{5.7}$$

$$\dot{z} = \frac{1}{\eta}\left[\dot{\boldsymbol{u}} - \nu\left(\beta|\dot{\boldsymbol{u}}||z|^{n-1}z + \gamma\dot{\boldsymbol{u}}|z|^{n}\right)\right] \tag{5.8}$$

$$\nu = 1 + \delta_\nu \varepsilon \tag{5.9}$$

$$\eta = 1 + \delta_\eta \varepsilon \tag{5.10}$$

$$\varepsilon = (1 - \alpha)k_0 \int_0^T \dot{\boldsymbol{u}} z dt \tag{5.11}$$

where, R ($\boldsymbol{u}$, z) is the restoring force. k_0 is the initial elastic stiffness. When other parameters remain unchanged, the increase of k_0 will increase the stiffness of the system, reduce the deformation of the system and the energy consumption. α ratio of the post-yield stiffness to the initial stiffness; z is the hysteretic variable; v is the intensity strengthening function; η is the stiffness degradation function; δ_v is the intensity enhancement rate; δ_η is the stiffness degradation rate; ε the cumulative hysteretic energy dissipation in the duration T, and β, γ, n are parameters influencing the hysteresis loop shape.

If β and γ are taken at different values, the system restoring force has soft or hard properties, that is, two parameters affect the nonlinear stiffness and nonlinear damping characteristics of the system. When $\beta > 0$, $\gamma > 0$, the system showed soft properties. When $\beta < 0$, $\gamma < 0$, the system showed hard characteristics. When the β/γ ratio is higher, the hysteretic curve envelope area is larger, and the system consumes more energy. The opposite was true when the β/γ ratio was small. When n is small, the curve is relatively smooth and the slip section of SSRCP part is longer. The larger n is, the shorter the partial slip segment is. When n approaches infinity, the smooth hysteretic curve degenerates into piecewise linear hysteretic curve.

The hysteretic curve of SSRCP shows weak asymmetry due to internal and external pressure and different material properties of each layer. In view of this, a shape control parameter δ describing the asymmetry is introduced in Eq. 5.5.

$$\dot{z} = \frac{1}{\eta}\left\{\dot{\boldsymbol{u}} - v\left[\beta \operatorname{sgn}(z\dot{\boldsymbol{u}}) + \delta\phi_1 + \gamma\right]\dot{\boldsymbol{u}}|z|^n\right\} \tag{5.12}$$

where, $\phi_1 = \operatorname{sgn}(z) + \operatorname{sgn}(\dot{\boldsymbol{u}})$.

Divide both sides of Eq. 5.12 by $\dot{u}$, and the derivative of z with respect to u can be obtained:

$$\frac{dz}{du} = \frac{1}{\eta}\left\{1 - v\left[\beta \operatorname{sgn}(z\dot{\boldsymbol{u}}) + \delta\phi_1 + \gamma\right]|z|^n\right\} \tag{5.13}$$

The asymmetry of peak restoring forces in positive and negative directions can be controlled by parameters $\pm 2\delta$.

However, the improved Bouc-Wen model is highly nonlinear differential equation. For the convenience of calculation, through the analysis of the mechanical response of SSRCP low-cycle cyclic loading, the whole loading process is divided into four stages: forward loading, forward unloading, reverse loading and reverse unloading. When loading forward, $z > 0$, $\dot{u} > 0$; when forward unloading, $z > 0$, $\dot{u} < 0$; when the load is reversed, $z < 0$, $\dot{u} > 0$; when reverse unloading occurs, $z < 0$, $\dot{u} < 0$. Thus, the absolute value sign in the equation can be eliminated. At this point, the improved hysteretic displacement Eq. 5.13 can be rewritten as:

$$\begin{cases} \frac{dz_1}{du_1} = \frac{1}{\eta}\left[1 - v(\beta + 2\delta + \gamma)z_1^n\right] & z > 0, \dot{u} > 0 \\ \frac{dz_1}{du_1} = \frac{1}{\eta}\left[1 - v(-\beta + \gamma)z_2^n\right] & z > 0, \dot{u} < 0 \\ \frac{dz_3}{du_3} = \frac{1}{\eta}\left[1 - v(-\beta + \gamma)z_3^n\right] & z < 0, \dot{u} > 0 \\ \frac{dz_4}{du_4} = \frac{1}{\eta}\left[1 - v(\beta - 2\delta + \gamma)z_4^n\right] & z < 0, \dot{u} < 0 \end{cases} \tag{5.14}$$

According to Eq. 5.11, the hysteretic energy at different stages in a period is different integral constant. Therefore, ε can be taken as a parameter of the improved hysteresis equation, that is, ε_i ($i = 1, 2, 3, 4$). At the same time, Eqs. 5.9–5.10 are substituted into Eq. 5.14 to obtain:

$$\begin{cases} \frac{dz_1}{du_1} = \frac{1}{1+\delta_\eta\varepsilon_1}\left[1 - (1+\delta_v\varepsilon_1)(\beta + 2\delta + \gamma)z_1^n\right], & z > 0, \dot{u} > 0 \\ \frac{dz_2}{du_2} = \frac{1}{1+\delta_\eta\varepsilon_2}\left[1 - (1+\delta_v\varepsilon_2)(-\beta + \gamma)z_2^n\right], & z > 0, \dot{u} < 0 \\ \frac{dz_3}{du_3} = \frac{1}{1+\delta_\eta\varepsilon_3}\left[1 - (1+\delta_v\varepsilon_3)(-\beta + \gamma)z_3^n\right], & z < 0, \dot{u} > 0 \\ \frac{dz_4}{du_4} = \frac{1}{1+\delta_\eta\varepsilon_4}\left[1 - (1+\delta_v\varepsilon_4)(\beta - 2\delta + \gamma)z_4^n\right], & z < 0, \dot{u} < 0 \end{cases} \tag{5.15}$$

The improved BW model has 12 model parameters, including 6 shape parameters (k_0, α, β, γ, δ, n), 1 regression parameters (δ_η), 1 strength enhancement parameter (δ_v), and 4 hysteretic energy parameters. These model parameters not only have unclear physical meanings, but also lack quantitative calculation formulas [21]. Therefore, it is necessary to identify the parameters of the asymmetric Bouc-Wen model according to the finite element simulation data of quasi-static reciprocating loading of SSRCP. By giving the initial conditions, Runge–Kutta method is used to solve ordinary differential equations.

5.5.2.2 Parameter Identification

The goal of parameter identification of the improved BW model is to find a set of model parameter vector Q and substitute it into the improved BW model to minimize the error between it and the experimental data. The parameter identification of the calculated restoring force model can be described as a nonlinear optimization problem:

$$\begin{cases} \min obj(\hat{Q}) \\ s.t. \hat{Q}_L \le \hat{Q} \le \hat{Q}_U \end{cases} \tag{5.16}$$

where, $\hat{Q}_L$ and $\hat{Q}_U$ are the upper and lower bounds of model parameter vectors $\hat{Q}$ respectively; $obj(\hat{Q})$ is the fitness function, which reflects the error between the measured value and the calculated value and is defined as:

$$obj(\hat{Q}) = \frac{1}{N}\sum_{i=1}^{N}\left[R(t_i) - R\left(t_i|\hat{Q}\right)\right] \tag{5.17}$$

where, N is the number of loading steps. $R(t_i)$ is the finite element simulation restoring force of SSRCP at t_i step. $R\left(t_i|\hat{Q}\right)$ is the restoring force at t_i step is calculated by substituting model parameters into the improved BW model.

As a stochastic optimization technique, particle swarm optimization (PSO) [22] iteratively updates particle positions by leveraging the particles' ability to retain and exchange information to converge toward the optimal solution. Owing to its simplicity and rapid convergence, the PSO algorithm has been extensively applied across various domains, including multi-objective optimization [23], pattern recognition [24], signal processing [25], damage detection [26], and the identification of hysteretic model parameters [27]. To enhance PSO's performance, several studies have integrated it with other intelligent algorithms [28, 29], introduced novel mechanisms into the standard PSO framework [30], or modified the inertia weight [31].

Optimization algorithm is an iterative process inspired by nature. The algorithm requires the function to be evaluated in a specific search space. In PSO algorithm, a group of particles called potential solutions is sent to the search space to determine their optimal location. For m-dimensional problems, the position and velocity of the i_{th} particle can be expressed as $X_i = (x_{i1}, x_{i2}, \ldots, x_{im})$ and $V_i = (v_{i1}, v_{i2}, \ldots, v_{im})$. According to the fitness function value, the optimal locations of particles in all particle species are recorded and stored in $P_g = (p_{g1}, p_{g2}, \ldots, p_{gm})$. The velocity and position of particles are expressed in Eqs. 5.18–5.19 respectively:

$$V_{id} = \omega V_{id} + c_1 \times \text{rand}_1 \times (P_{id} - X_{id}) + c_2 \times \text{rand}_2 \times \left(P_{gd} - X_{id}\right) \tag{5.18}$$

$$X_{id} = X_{id} + V_{id} \tag{5.19}$$

where ω is inertia coefficient; c_1 and c_2 are acceleration coefficients; and $rand_i$ ($i = 1, 2$) random number.

Inertia and acceleration coefficients are calculated using the method proposed by Clerc and Kennedy [32]. First, two construction coefficients are calculated using

$$\chi = \frac{2\kappa}{\left|2 - \zeta - \sqrt{\zeta^2 - 4\zeta}\right|} \tag{5.20}$$

$$\zeta = \zeta_1 + \zeta_2 \tag{5.21}$$

The value of ζ_1 and ζ_2 should be such that $\zeta > 4$,

Table 5.2 Parameter identification results under different algorithms

	PSO	GA	SA
Fitness	184.725	199.813	423.736

$$\omega = \chi \tag{5.22}$$

$$c_1 = \chi \zeta_1 \tag{5.23}$$

$$c_2 = \chi \zeta_2 \tag{5.24}$$

To assess the accuracy and efficiency of the PSO algorithm in parameter identification, we compared it with the GA and SA for identifying 12 parameters of a nonlinear restoring force model under an excitation range of − 0.35 rad to 0.35 rad. During the loading process, the population size for all three algorithms was set to 50, with a maximum of 200,000 iterations. GA used a crossover probability of 0.7 and a mutation probability of 0.1, applying a roulette selection mechanism, while the SA parameters followed the default MATLAB settings.

The final fitness function values for each algorithm, shown in Table 5.2, indicate that PSO achieved the lowest value, with GA slightly higher and SA considerably higher. The identified parameters were incorporated into the improved BW model and validated against finite element simulation data (Fig. 5.30). Results showed that both PSO and GA converged closely to the true values, with PSO exhibiting the best consistency. In contrast, SA produced significantly larger errors.

Figure 5.31 illustrates the convergence curves of the fitness function values for the three algorithms, clearly demonstrating their accuracy and efficiency. Initially, GA exhibits strong search capabilities and achieves convergence first, indicating a high

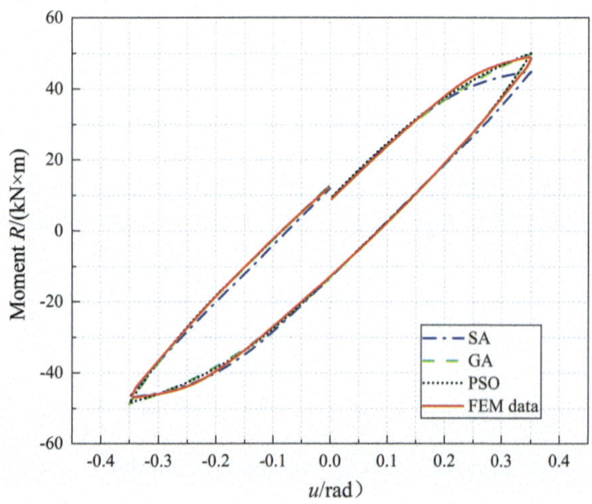

Fig. 5.30 Hysteresis curve under different algorithms

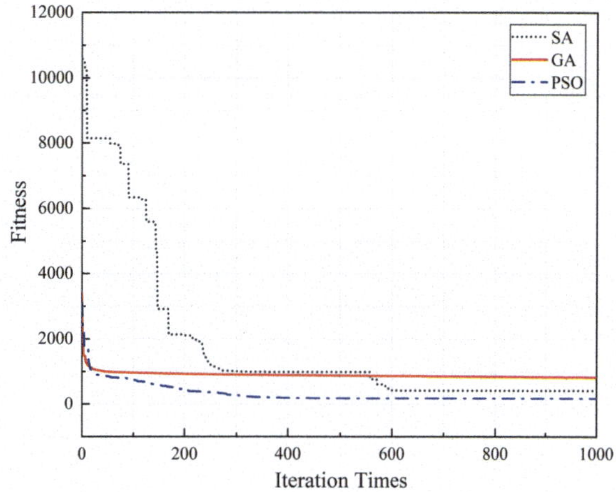

Fig. 5.31 Convergence curves different algorithms fitness function values under

convergence speed. However, it tends to suffer from premature convergence, often getting trapped in local optima, leading to greater errors in parameter identification. SA shows a more gradual, step-wise convergence, reaching stability after approximately 600 iterations, which is significantly slower than PSO's 350 iterations. Thus, PSO proves to be the most efficient, offering rapid and accurate identification of the improved Bouc-Wen model parameters.

The parameter identification results of improved BW model obtained by PSO algorithm, GA and SA were compared, as shown in Fig. 5.32. It can be seen that the recognition effects of k_0, n, α, β, δ, δ_η and δ_v are roughly the same for PSO and GA. Compared to the other two algorithms, the PSO algorithm enhances the accuracy of parameter identification and significantly reduces fluctuations during the convergence process. Parameters β and γ, which control the shape of the hysteresis loop, lack direct physical significance. Different combinations of these parameters can yield the same model output under identical inputs, resulting in non-unique solutions to the inverse problem. The algorithm's performance plays a key role in determining the accuracy of parameter identification, making the results generally applicable. Therefore, even when the true values of the parameters in the improved Bouc-Wen model change, the recognition patterns of the three algorithms remain consistent under the same conditions. Due to space constraints, results for other parameter configurations are not provided.

5.6 Conclusions

Hysteretic Behavior and Stiffness: SSRCP exhibits distinct hysteretic behavior under cyclic bending loading, with its overall stiffness progressively decreasing over successive loading cycles. Key factors such as internal pressure, diameter-to-thickness ratio, and the spiral angle of the steel strip significantly impact the

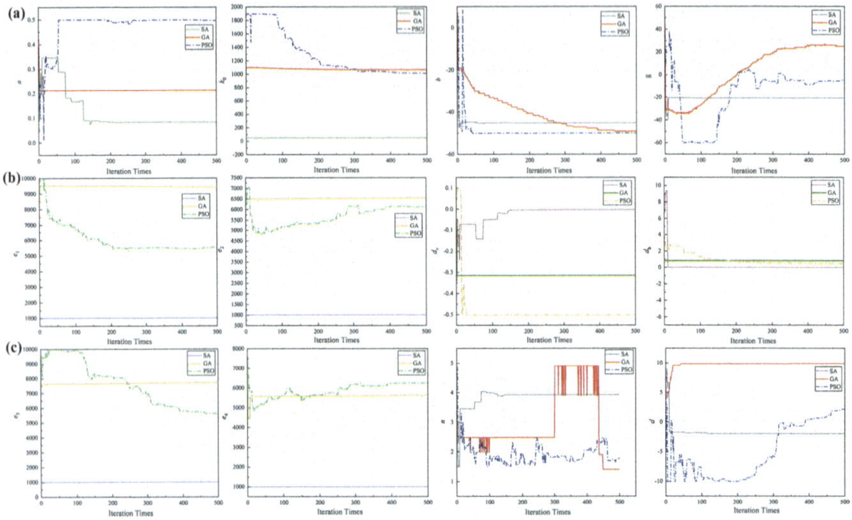

Fig. 5.32 Parameter identification results under different algorithms

stiffness and flexibility of SSRCP. The frictional energy dissipation in SSRCP increases rapidly during low-cycle cyclic loading, influenced by internal pressure, diameter-thickness ratio, and friction coefficient.

Modeling and Parameter Identification: The chapter introduces an improved Bouc-Wen model to describe the hysteretic properties of SSRCP, incorporating asymmetric factors and intensity enhancement factors. The model divides the loading process into four sections, simplifying the equation-solving process. Parameter identification conducted using stochastic intelligent algorithms, including PSO, GA, and SA, reveals that PSO achieves superior accuracy and efficiency. The improved model provides a reliable theoretical basis for describing SSRCP's hysteretic properties, facilitating its integration into commercial software for practical applications.

In summary, this chapter provides a comprehensive understanding of SSRCP's behavior under various loading conditions, highlighting its flexibility, load-bearing capacity, and hysteretic behavior. The improved Bouc-Wen model offers a robust tool for accurately describing these characteristics, supported by efficient parameter identification methods.

This chapter is adapted from the article 'Nonlinear hysteretic behavior of wound reinforced composite pipe with steel strip in service,' originally published in *Ocean Engineering* (2022). DOI: https://doi.org/10.1016/j.oceaneng.2021.110302, and 'Nonlinear hysteretic modeling and parameter identification of steel strip winding reinforced composite pipe', originally published in *Ocean Engineering* (2021). Reproduced with permission from Elsevier.

References

1. Shuai Y (2017) Ultimate capacity study for unbonded flexible pipes under pressures. Zhejiang University, Hangzhou
2. He Q, Wang P (2017) Current situation and prospect of deep-sea energy development. Ocean Development and Management 34(12):18–23
3. Yang X, Liping S, Shangmao A (2013) Recent research of deepwater unbonded flexible pipe tensile armor layer's buckling behaviors. Ocean Eng 01:99–106
4. Ting L (2018) Mechanical behaviors and reliability analysis of steel strip reinforced flexible pipes. Zhejiang University, Hangzhou
5. Bai Y, Liu T, Ruan W et al (2017) Mechanical behavior of metallic strip flexible pipe subjected to tension. Compos Struct 170(Complete):1–10. https://doi.org/10.1016/j.compstruct.2017.02.044
6. Bhardwaj U, Teixeira AP, Guedes Soares C (2020) Reliability assessment of a subsea pipe-in-pipe system for major failure modes. Int J Press Vessels Pip 188:104177. https://doi.org/10.1016/j.ijpvp.2020.104177
7. Paris: IEA (2022) IEA. Electricity market report. https://www.iea.org/reports/electricity-market-report-july-2022
8. Chatzi EN, Smyth AW, Masri SF (2010) Experimental application of on-line parametric identification for nonlinear hysteretic systems with model uncertainty. Struct Saf 32(5):326–337. https://doi.org/10.1016/j.strusafe.2010.03.008
9. Caughey TK (1960) Sinusoidal excitation of a system with bilinear hysteresis. 640–643
10. Iwan WD (1966) A distributed-element model for hysteresis and its steady-state dynamic response. 893–900
11. Bouc R (1971) A mathematical model for hysteresis. Acta Acust Acust 24(1):16–25
12. Masri SF (1975) Forced vibration of the damped bilinear hysteretic oscillator. J Acoust Soc Am 57(1):106–112
13. Baber TT, Wen Y-K (1981) Random vibration of hysteretic, degrading systems. J Eng Mech Div 107(6):1069–1087
14. Jayakumar P, Beck JL (1987) Proceedings structural safety evaluation on system identification approaches. Lambracht, Germany, p 82
15. Vaiana N et al (2018) A class of uniaxial phenomenological models for simulating hysteretic phenomena in rate-independent mechanical systems and materials. Nonlinear Dyn 93:1647–1669. https://doi.org/10.1007/s11071-018-4282-2
16. Bouc R (1967) Forced vibrations of mechanical systems with hysteresis. In: Proceedings of the fourth conference on nonlinear oscillations, Prague, p 315
17. Wen Y-K (1976) Method for random vibration of hysteretic systems. J Eng Mech Div 102(2):249–263
18. Baber TT, Noori MN (1985) Random vibration of degrading, pinching systems. J Eng Mech 111(8):1010–1026
19. Kwok NM et al (2007) Bouc-Wen model parameter identification for a MR fluid damper using computationally efficient GA. ISA Trans 46(2):167–179. https://doi.org/10.1016/j.isatra.2006.08.005
20. Zhu W, Wang D-H (2012) Non-symmetrical Bouc-Wen model for piezoelectric ceramic actuators. Sens Actuators, A 181:51–60. https://doi.org/10.1016/j.sna.2012.03.048
21. Smyth AW et al (1999) On-line parametric identification of MDOF nonlinear hysteretic systems. J Eng Mech 125(2):133–142. https://doi.org/10.1061/(ASCE)0733-9399(1999)125:2(133)
22. Kennedy J, Eberhart R (1995) Particle swarm optimization. In: Proceedings of ICNN'95–International conference on neural networks, vol 4. IEEE, pp 1942–1448
23. Parsopoulos KE, Vrahatis MN (2002) Particle swarm optimization method in multiobjective problems. In: Proceedings of the 2002 ACM symposium on applied computing

24. Liu L, Liu W, Cartes DA (2008) Particle swarm optimization-based parameter identification applied to permanent magnet synchronous motors. Eng Appl Artif Intell 21(7):1092–1100. https://doi.org/10.1016/j.engappai.2007.10.002
25. Lei B et al (2015) Optimal and secure audio watermarking scheme based on self-adaptive particle swarm optimization and quaternion wavelet transform. Signal Process 113:80–94. https://doi.org/10.1016/j.sigpro.2014.11.007
26. Wang D et al (2015) Damage identification in shear-type structures using a proper orthogonal decomposition approach. J Sound Vib 355:135–149. https://doi.org/10.1016/j.jsv.2015.06.043
27. Vaiana N, Losanno D, Ravichandran N (2021) A novel family of multiple springs models suitable for biaxial rate-independent hysteretic behavior. Comput Struct 244:106403. https://doi.org/10.1016/j.compstruc.2020.106403
28. Guan F, Dai L, Xia M (2014) Pretension optimization and verification test of double-ring deployable cable net antenna based on improved PSO. Aerosp Sci Technol 32(1):19–25. https://doi.org/10.1016/j.ast.2013.11.008
29. Raouf N, Pourtakdoust SH (2015) Launch vehicle multi-objective reliability-redundancy optimization using a hybrid genetic algorithm-particle swarm optimization. Proc Inst Mech Eng Part G: J Aerospace Eng 229(10):1785–1797. https://doi.org/10.1177/0954410014560381
30. Sun J, Liu X (2013) A novel APSO-aided maximum likelihood identification method for Hammerstein systems. Nonlinear Dyn 73:449–462. https://doi.org/10.1007/s11071-013-0800-4
31. Taherkhani M, Safabakhsh R (2016) A novel stability-based adaptive inertia weight for particle swarm optimization. Appl Soft Comput 38:281–295. https://doi.org/10.1016/j.asoc.2015.10.004
32. Wang L, Tang Y, Ma T, Zhong J, Li Z, Zhang Y et al (2020) Stress concentration analysis of butt welds with variable wall thickness of spanning pipelines caused by additional loads. Int J Press Vessels Pip 182:104075. https://doi.org/10.1016/j.ijpvp.2020.104075

Chapter 6
Buckling Failure Mechanism of Liner Pipe in Bimetal Mechanical Clad Pipes

6.1 Introduction

Bimetal mechanical clad pipes, crucial in combating corrosion in oil, gas, and marine applications, consist of a durable outer pipe made of low alloy or carbon steel, and a corrosion-resistant inner liner of alloy or stainless steel [1]. The unique construction, enhanced by mechanical or metallurgical bonding techniques, offers superior performance compared to single-material pipes, though it presents specific mechanical challenges, particularly concerning the liner pipe, crucial for the clad pipe's safety. Buckling failure in the liner pipe, attributed to various stresses and material interactions, poses a significant threat, potentially leading to leaks, ruptures, and severe economic and environmental consequences [2]. This chapter aims to explore the buckling failure mechanism in the liner pipe of bimetal mechanical clad pipes, examining contributing factors and impacts on pipe integrity to improve the reliability of these essential components in the oil and gas industry.

6.2 Hydroforming Principle

Several methods exist for the formation of bimetal mechanical clad pipes, among which hydroforming is particularly prevalent. Figure 6.1 shows the schematic diagram of the hydroforming process used for the bimetallic mechanical clad pipe. The hydroforming principle involves inserting a liner pipe within an outer pipe, maintaining an initial gap between them. The process starts by applying uniform internal pressure to the inner wall of the liner pipe, causing it to undergo initial elastic deformation. As the internal pressure increases, the outer wall of the liner pipe contacts the inner wall of the outer pipe. Subsequently, both pipes expand concurrently under the influence of this internal pressure. During this phase, the liner pipe experiences plastic deformation, while the outer pipe undergoes elastic deformation. Upon the

H. Zhang et al., *Submarine Pipeline Integrity: Assessment of Failure Modes and Advanced Evaluation Techniques*, Ocean Engineering & Oceanography 28,
https://doi.org/10.1007/978-3-031-92092-9_6

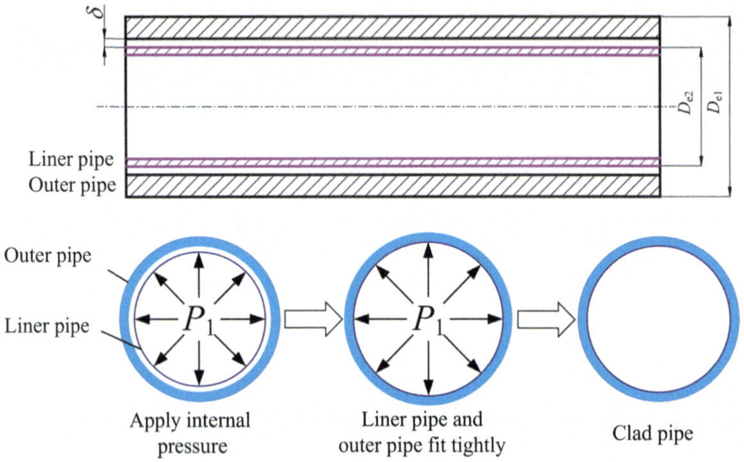

Fig. 6.1 Hydroforming process of a bimetal clad pipe

release of the internal pressure, the liner pipe exhibits less rebound compared to the outer pipe. This differential in rebound results in an interference fit and residual pressure between the two pipes, effectively bonding them together. This method ensures a robust and durable bond, essential for the integrity and functionality of the clad pipe.

The fabrication process of bimetallic clad pipes necessitates the careful management of residual stress between the liner and outer pipes. It is crucial to maintain this residual stress within an optimal range; excessive residual stress can lead to buckling of the liner pipe during the forming process. The minimum and maximum forming pressures are critical parameters in this context and are quantitatively defined by Eqs. 6.1 and 6.2, respectively [3]. These equations serve as guidelines to ensure the structural integrity and reliability of the clad pipe throughout the forming process.

$$P_{\min} \geq \left(\frac{K_1^2 - 1}{2K_1^2} + \text{In}K_2\right)\sigma_{y2} \tag{6.1}$$

$$P_{\max} = \frac{K_1^2 - 1}{2K_1^2}\sigma_{y1} + \sigma'_{y2}\text{In}K_2 \tag{6.2}$$

$$\sigma'_{y2} = \sigma_{y2} + E'_i(2\delta/D_1) \tag{6.3}$$

where $K_1 = D_1/D_2$, $K_2 = d_1/d_2$, σ_{y1} is the yield strength of outer pipe, σ_{y2} is the yield strength of liner pipe, D_1 is the outer diameter of outer pipe, D_2 is the outer diameter of liner pipe, d_1 is the average diameter of outer pipe, d_2 is the average diameter of liner pipe, σ'_{y2} is the strain strengthening stress of liner pipe, δ is the clearance between inner and outer pipes and E'_i is the elastic modulus.

Upon calculation, the maximum forming internal pressure P_{max} of the bimetallic clad pipe is 55.4 MPa, and the minimum forming internal pressure P_{min} is 39.9 MPa. The maximum and minimum residual circumferential stress of the clad pipe are 165 and 108 MPa, respectively. Additionally, the maximum and minimum residual contact pressures are 2.72 and 0.65 MPa, respectively. This data indicates a proportional relationship between the forming pressure and the resultant stresses: as the forming pressure increases, there is a corresponding increase in both the residual circumferential stress and the residual contact pressure within the bimetallic clad pipe.

6.3 Numerical Simulation Model

The forming process of bimetallic clad pipes can be effectively simulated using the advanced finite element analysis software ABAQUS. the outer pipe is composed of X65 steel, while the liner pipe is fabricated from SS304 stainless steel. The length of the bimetallic clad pipe is 0.3 m. The material constitutive model adopts the Ramberg–Osgood model, modified by Hill, as shown in Eq. 6.4.

$$\varepsilon = \frac{\sigma}{E} + 0.002\left(\frac{\sigma}{\sigma_y}\right)^n \tag{6.4}$$

where ε is the total strain, E is the initial elastic modulus and σ_y is the yield strength. The specific geometric and material parameters are shown in Table 6.1 [4].

In the simulation model, the outer pipe is meshed with hexahedral solid elements, while the liner pipe employs shell elements. The element count is 6400 for the outer pipe and 21,988 for the liner pipe, ensuring a mesh that is both adequate and free from irregularities. The contact interaction is modeled using a surface-to-surface contact, employing hard contact with a limited-slip method to accurately simulate the physical behavior at the interface between the two pipes.

The boundary conditions for the numerical model are depicted in Fig. 6.2. At the model's left end, symmetry constraints are applied, while the right end is configured to restrict axial movement. The hydraulic forming process is modeled in accordance with the experimental steps outlined in the study by Zeng et al. [3]. The simulation process encompasses two primary stages: initially, hydraulic forming is conducted by

Table 6.1 Clad pipe's geometry and material parameters

Parameters	External diameter D (mm)	Wall thickness t (mm)	Diameter thickness ratio D/t	Elastic modulus E (GPa)	Poisson's ratio v	Yield strength σ_y (MPa)	Hardening exponent n
Outer pipe	325	14.3	22.6	210	0.3	465	16
Liner pipe	294.4	3	98.8	198	0.3	277	13

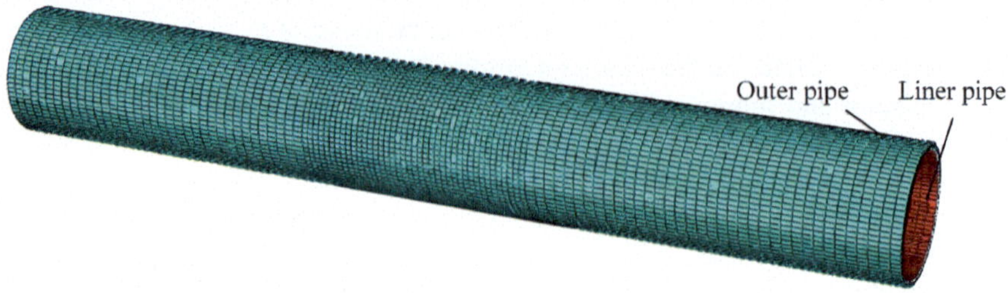

Fig. 6.2 Finite element of the bimetal clad pipeline

applying uniform internal pressure, ensuring adherence between the liner and outer base pipes, and inducing elastic deformation in the latter, which results in residual stress upon the release of pressure. Subsequently, the model undergoes a phase of pure bending. This approach ensures that the liner and outer pipes are in close contact, leading to the generation of residual stress in both pipes upon the release of the forming pressure. A Kinematic Coupling is created between a reference point and a node on the clad pipe's end face, enabling the application of a rotational angle at the reference point for bending moment loading. Subsequently, axial force, internal pressure, and external pressure are applied, facilitating the simulation of complex loading conditions.

Rotation angle $\theta = 0.25$ rad, forming internal pressure $P_c = 45$ MPa, external pressure $P_1 = 2$ MPa, internal pressure $P_2 = 1.5$ MPa, the wall thickness of outer pipe $t_1 = 15$ mm, the wall thickness of liner pipe $t_2 = 3$ mm, axial force $F_t = 200$ kN and liner and outer pipe clearance $\delta = 1$ mm are selected as the basic parameter.

The curvature is normalized by Eq. 6.5. The bending normalized curvature nominal value $K_E = k/k_N$. The calculation formula of the normalized ovality is shown in Eq. 6.6. The bending moment M is normalized by Eq. 6.7, so the normalized bending moment $m = M/M_1$.

$$k_N = t_1/d_1^2 \tag{6.5}$$

$$\zeta = (d_{max} - d_{min})/(d_{max} + d_{min}) \tag{6.6}$$

$$M_1 = \sigma_{y1} d_1^2 t_1 \tag{6.7}$$

where, t_1 is the thickness of the outer pipe, d_{max} is the maximum average diameter of the liner pipe, and d_{min} is the minimum average diameter of the liner pipe.

To validate the reliability of the simulation model, a numerical simulation consistent with the experiment described in Ref. [5] was established for comparison. Figure 6.3 shows a comparison of the residual contact stress between the numerical model, experimental data, and analytical solutions. The results of the three methods are closely aligned, with the maximum discrepancy compared to the experimental

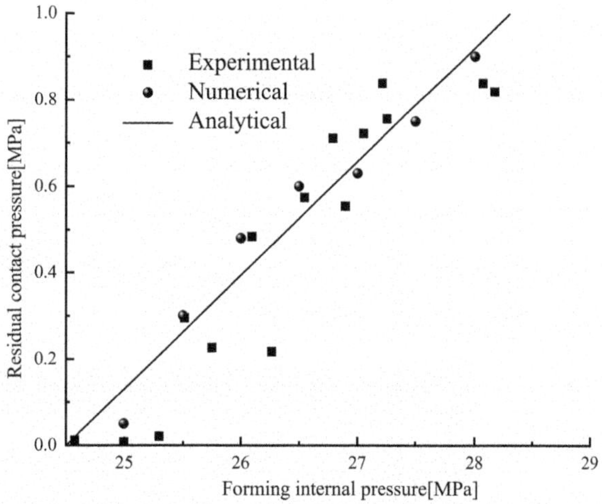

Fig. 6.3 Comparison of numerical model, experimental, and analytical model results of residual contact stress

data being 8.4%. The theoretical analytical solution exceeds the finite element numerical solution, mainly because it assumes the pipes are ideal cylinders, where contact or separation between the inner and outer pipes happens simultaneously. In contrast, the finite element model, with its non-uniform mesh node deformation, does not replicate this simultaneous action, making the numerical simulation more closely aligned with experimental values than the theoretical analysis.

6.4 Results and Discussions

6.4.1 Mechanical Behaviors

Figure 6.4a illustrates the stress distribution in the inner liner pipe under a bending load. At a bending moment of 1.37×10^5 N m, the composite pipe shows a uniform stress distribution, with lower stress at the ends and two high-stress regions in the middle. As the bending moment increases, the compression side's stress distribution becomes more pronounced, especially in the middle with alternating high and low-stress zones. At a bending moment of 3.49×10^5 N m, high stress appears in the middle with stress relief at the edges. When the moment reduces to 3.02×10^5 N m, the inner liner pipe buckles, forming an "X" shaped high-stress area in the middle.

Figure 6.4b depicts the equivalent plastic strain distribution in the inner liner pipe across different stages. In Stage I, under low bending moments, the pipe undergoes elastic deformation without separation between the inner and outer pipes. In Stage II, increased bending moments cause plastic deformation on the compression side of the inner pipe, mainly in the middle, while the ends remain elastic, leading to the initial separation of the pipes. Stage III sees further bending moment increases, concentrating plastic deformation in the middle of the compression side and more

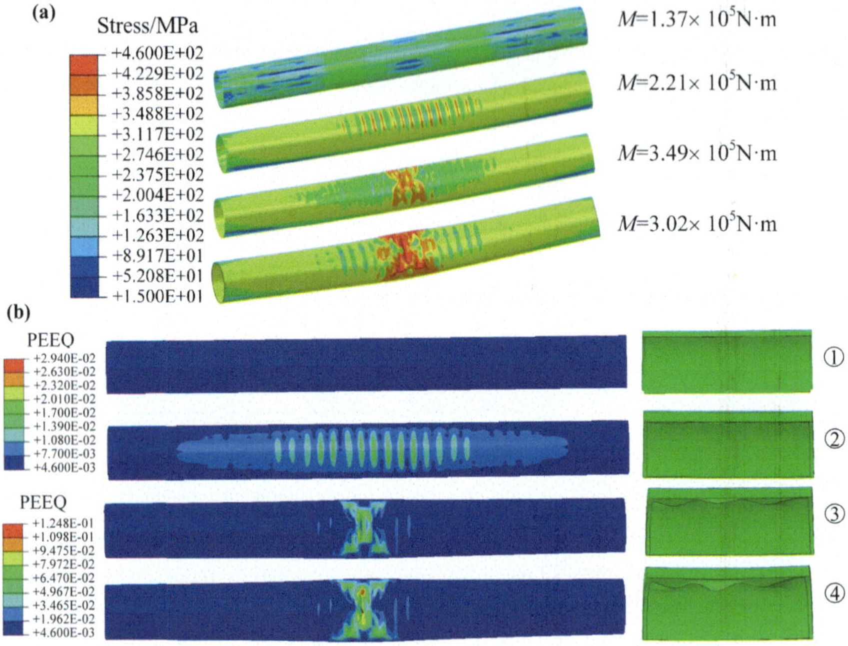

Fig. 6.4 Stress and equivalent strain distributions of the liner pipe under bending load

pronounced pipe separation, with the onset of wrinkling bifurcation. Finally, in Stage IV, the middle section's plastic deformation area expands, forming an "X" pattern, and the composite pipe shows clear wrinkling bifurcation.

Figure 6.5a illustrates the stress distribution in a liner pipe under various complex loadings, highlighting distinct stress patterns between the liner and outer pipes. At a bending moment of 1.34×10^5 N m, high-stress regions emerge at both ends of the pipes on the compression side, with low stress in the middle. As the bending moment increases to 2.07×10^5 N m, the entire compression side of the outer pipe experiences a significant stress increase, while the liner pipe only shows stress concentration in the middle. At 3.37×10^5 N m, the outer pipe's high stress diminishes, but stress concentration persists in the liner pipe's middle. When the bending moment decreases to 2.79×10^5 N m, the outer pipe avoids plastic deformation, but the liner pipe buckles, forming an '8-shaped' high-stress region in the middle, with stress gradually releasing at both ends.

Figure 6.5b presents the equivalent plastic strain distribution in the liner pipe under complex loadings. Initially, minor plastic deformation occurs in the liner pipe. As the bending moment rises, the compression side of the liner pipe undergoes plastic deformation, showing low equivalent plastic strain at both ends and high strain in the middle. Further increase in bending moment concentrates plastic deformation in the middle of the compression side, resulting in a circular and uniform strain distribution, while both ends experience strain release. The liner pipe starts buckling

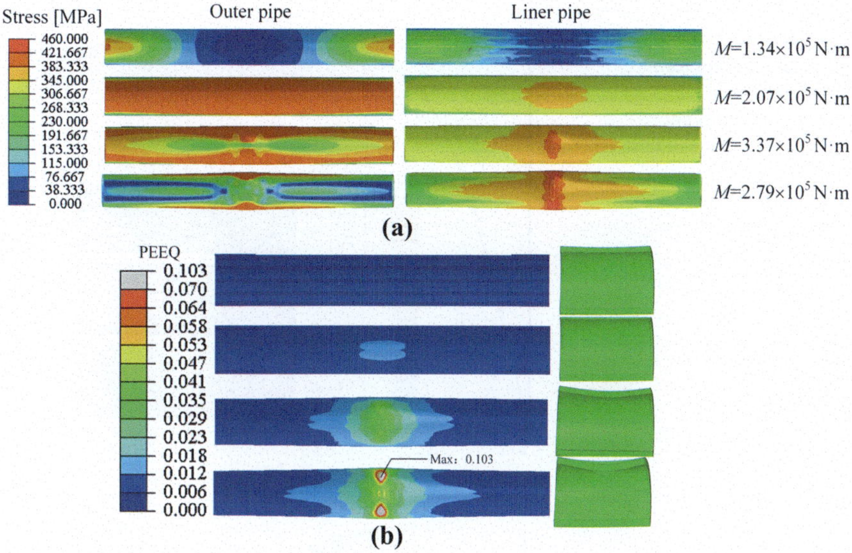

Fig. 6.5 Stress and equivalent strain distributions of the liner pipe under complex loadings. **a** Stress, **b** strain

from the middle, leading to separation from the outer pipe. Eventually, the middle part's plastic deformation intensifies, enlarging the high plastic strain region and causing two high-strain areas to form, resulting in wrinkles in the clad pipe.

6.4.2 Sensitivity Analysis

6.4.2.1 Modeling Process

In Fig. 6.6a, initially, the normalized bending moment increases linearly under load, indicating elastic deformation of the liner pipe (Stage I). As curvature increases, the bending moment rises slowly, with minimal impact from forming pressure. In Stage II, the liner pipe undergoes plastic deformation, causing a reduction in structural performance. In Stage III, the bending moment on the composite pipe decreases, and the liner pipe's equivalent plastic strain increases, indicating buckling and a sharp decline in performance. The point where the bending moment peaks and then decreases is the critical curvature.

In Fig. 6.6b, at low curvature, the wrinkle amplitude of the liner pipe remains constant, indicating no separation from the outer base pipe. As normalized curvature increases, at $K_E = 0.28$, the normalized fold amplitude suddenly enlarges, marking the onset of folding and separation between the pipes, and a decrease in the liner pipe's buckling resistance. Higher forming internal pressures result in lower normalized fold

Fig. 6.6 Mechanical responses of the bimetal clad pipes with different forming internal pressures. **a** Normalized bending moment, **b** normalized fold amplitude

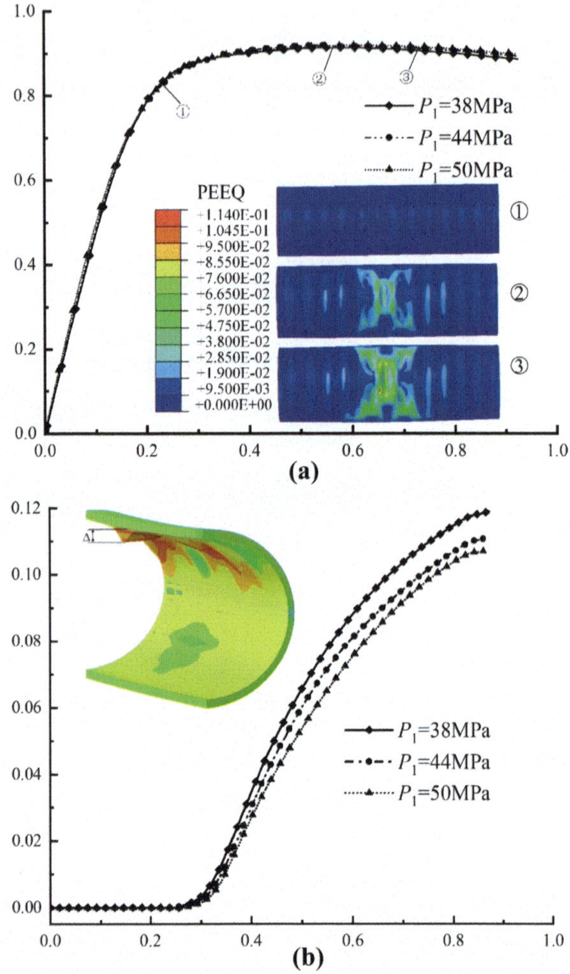

amplitudes at the same curvature, indicating stronger buckling resistance of the liner pipe.

Figure 6.7 displays the clad pipe's mechanical behavior under varying forming internal pressures, with selected residual stresses at 40 MPa, 45 MPa, 50 MPa, and 55 MPa, resulting in 108 MPa, 126 MPa, 143 MPa, and 165 MPa respectively. In Fig. 6.7a, the clad pipe's normalized bending moment under complex loading initially rises almost linearly, indicating elastic deformation of the liner pipe. Upon reaching the maximum bending moment, the corresponding curvature is termed critical curvature. As forming internal pressure increases, so does the critical curvature needed to reach the maximum bending moment, showcasing enhanced resistance of the clad pipe to bending moments with higher forming pressures.

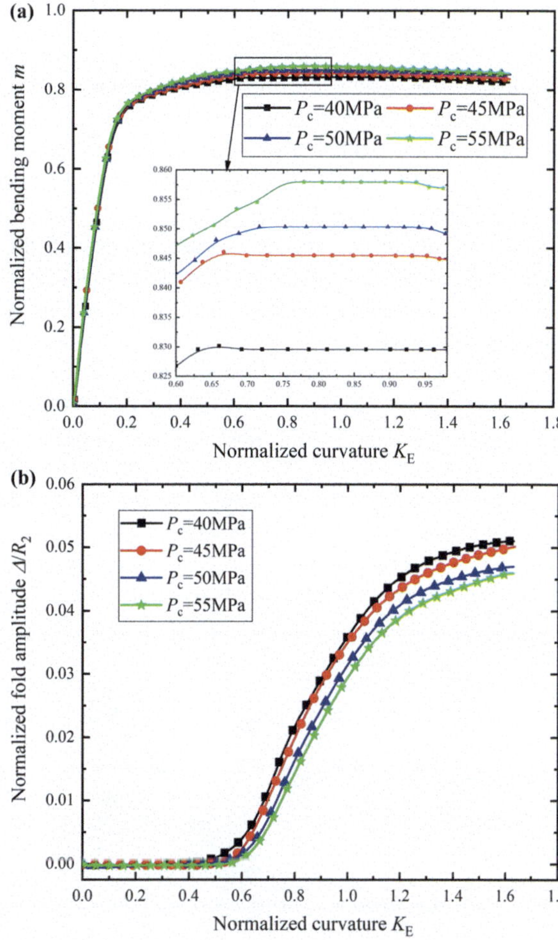

Fig. 6.7 Mechanical responses of the bimetal clad pipes under complex loading conditions at varying forming internal pressures. **a** Normalized bending moment, **b** normalized fold amplitude

In Fig. 6.7b, the normalized fold amplitude remains constant when $K_E < 0.5$, signifying no separation between the liner and outer pipes. However, as the outer pipe's wall thickness increases, the initial normalized curvature of the curve also rises, suggesting that higher forming internal pressures postpone the onset of liner pipe folding. Additionally, under the same normalized curvature, a higher forming internal pressure results in a lower normalized fold amplitude for the liner pipe, indicating that a thicker wall from higher forming internal pressure enhances the clad pipe's buckling resistance.

6.4.2.2 Working Pressure

Figure 6.8 shows how the mechanical response of composite pipes varies with internal working pressure. Figure 6.8a shows that higher internal pressures lead to less plastic strain and fewer wrinkles in the liner pipe, reducing its plastic deformation. In Fig. 6.8b, increased internal pressure enhances the composite pipe's resistance to bending moments. Figure 6.8c reveals that higher internal pressures result in lower ovality at the same curvature, improving the liner pipe's buckling resistance. Figure 6.8d indicates that without internal pressure, the composite pipe wrinkles earlier, but higher pressures lead to lower fold amplitudes, signifying greater buckling resistance.

Figure 6.9 displays the stress distribution of the liner pipe section under varying internal pressures, showing a symmetric distribution about the 90°–270° vertical line. The pipe section deforms into an ellipse under complex loading, with maximum stresses shifting from 12°–28° and 152°–178° at 0.5 MPa to 26°–45° and 135°–162° at higher pressures, indicating a delay in buckling failure and a reduction in buckling degree with increased internal pressure.

Figure 6.10 presents the equivalent plastic strain distribution of the liner pipe section, also symmetrically distributed and showcasing a "bat-shaped" pattern. The

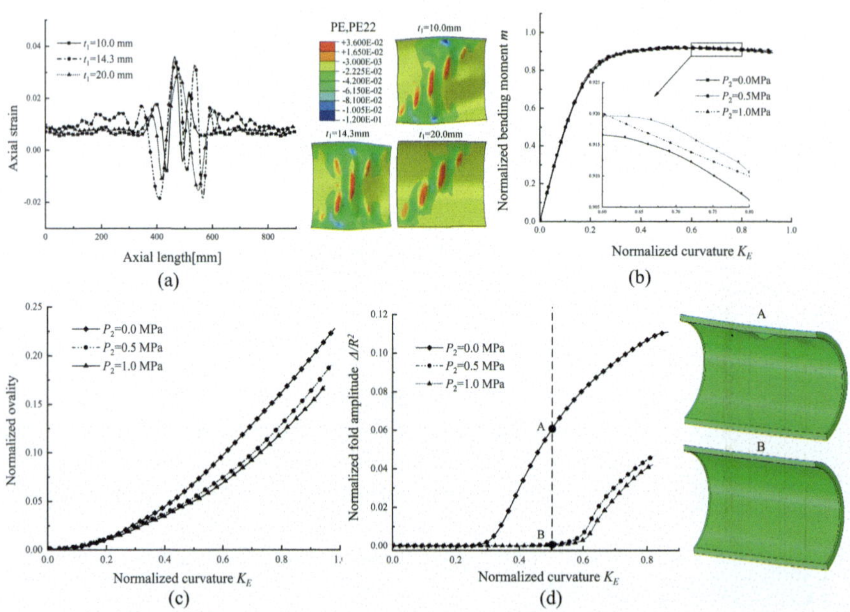

Fig. 6.8 Mechanical response of clad pipe with different internal working pressures **a** Plastic strain curve and plastic strain contour plot of the liner pipe. **b** Relationship between bending moment and curvature, **c** relationship between ovality rate of composite pipe and curvature, **d** relationship between normalized wrinkle amplitude and curvature

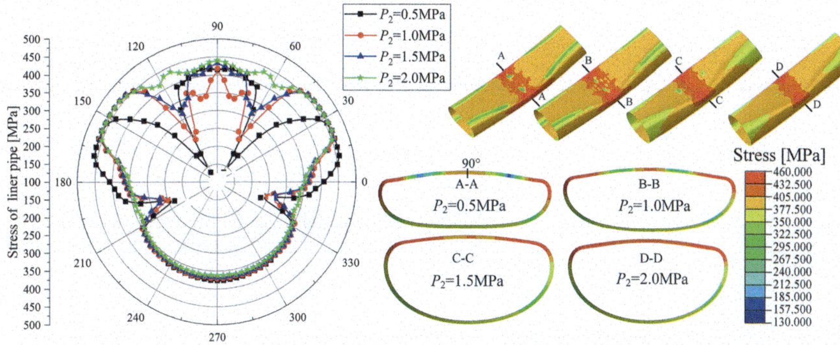

Fig. 6.9 Stress distribution curve of liner pipe section under different internal pressures

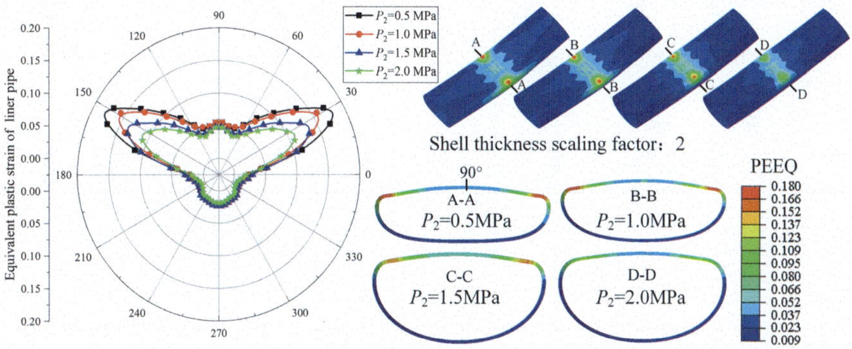

Fig. 6.10 Equivalent plastic strain distribution of liner pipe section under different internal pressures

maximum strain occurs at 28° and 152°, decreasing with rising internal pressure, which helps in delaying the buckling failure.

Figure 6.11 illustrates the axial plastic strain distribution on the compression side of the liner pipe, with high strain covering the entire compression side at 0.5 MPa and concentrating in the middle at 2.0 MPa. The strain distribution is symmetric, and increasing internal pressure significantly reduces compressive strain, indicating enhanced buckling resistance.

Figure 6.12a shows the mechanical response of the clad pipe, with ovality initially increasing gradually, followed by a sharp rise in the growth rate beyond a certain point, indicating a weakening in buckling resistance. However, higher internal pressures result in lower ovality under the same conditions, showcasing improved buckling resistance. In Fig. 6.12 b, the fold amplitude remains constant up to a certain point, indicating no separation between the liner and outer pipes. Beyond this point, the buckling resistance of the liner pipe gradually diminishes; however, increasing internal pressure can delay the onset of wrinkling and enhance buckling resistance.

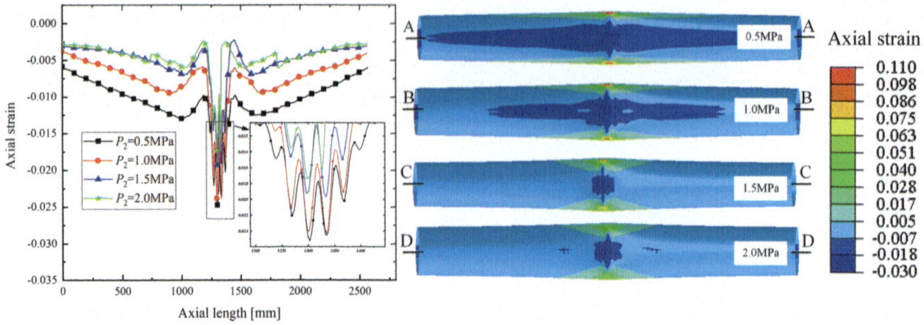

Fig. 6.11 Axial plastic strain distribution of the liner pipe under different internal pressures

Fig. 6.12 Mechanical response of clad pipe under different internal pressures. **a** Normalized bending moment, **b** normalized fold amplitude

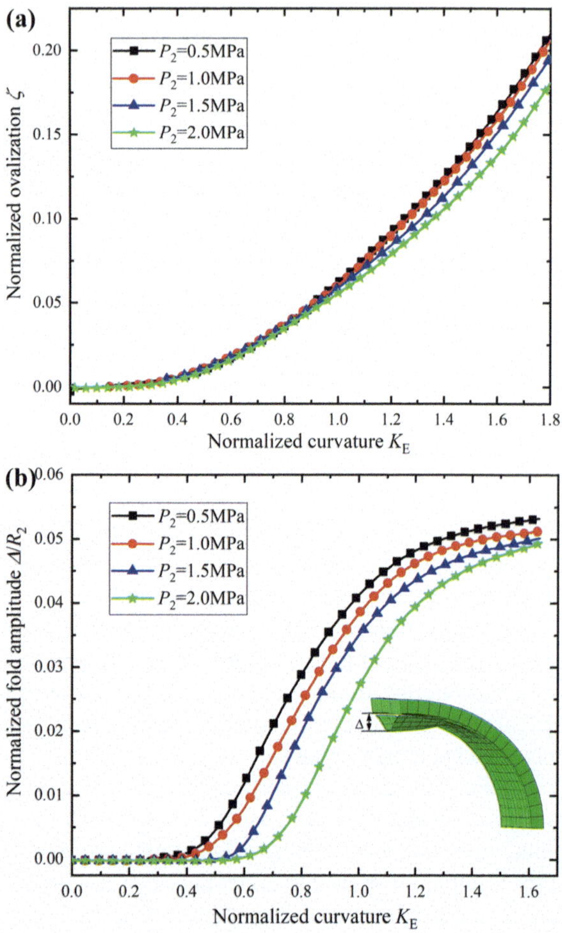

6.4.2.3 Liner Pipe Wall Thickness

Figure 6.13 demonstrates the influence of varying liner pipe wall thicknesses on composite pipe mechanics. In Fig. 6.13a, increased wall thickness correlates with reduced wrinkle count in the liner pipe. Figure 6.13b shows that thicker walls lead to higher bending moments at identical curvatures. Figure 6.13c reveals that greater wall thickness results in lower ellipticity at the same curvature, indicating enhanced buckling resistance. Finally, Fig. 6.13d indicates that increased wall thickness corresponds to lower normalized wrinkle amplitudes, further signifying improved buckling resistance.

Figure 6.14 illustrates how varying liner pipe wall thicknesses affect stress distribution. Thinner walls (2.5 and 3.0 mm) show maximum stresses at 15°–40° and 140°–175°, while a 3.5 mm wall shifts this to 25°–45° and 135°–165°, and a 4.0 mm wall centralizes maximum stress near 90°. As wall thickness increases, maximum stress decreases and spreads across the compression surface, delaying buckling failure.

Figure 6.15 presents a "bat-shaped" equivalent plastic strain distribution, peaking at 15°–25° and 145°–155°. Increased wall thickness shortens the bat wings, reducing maximum strain and enhancing buckling resistance.

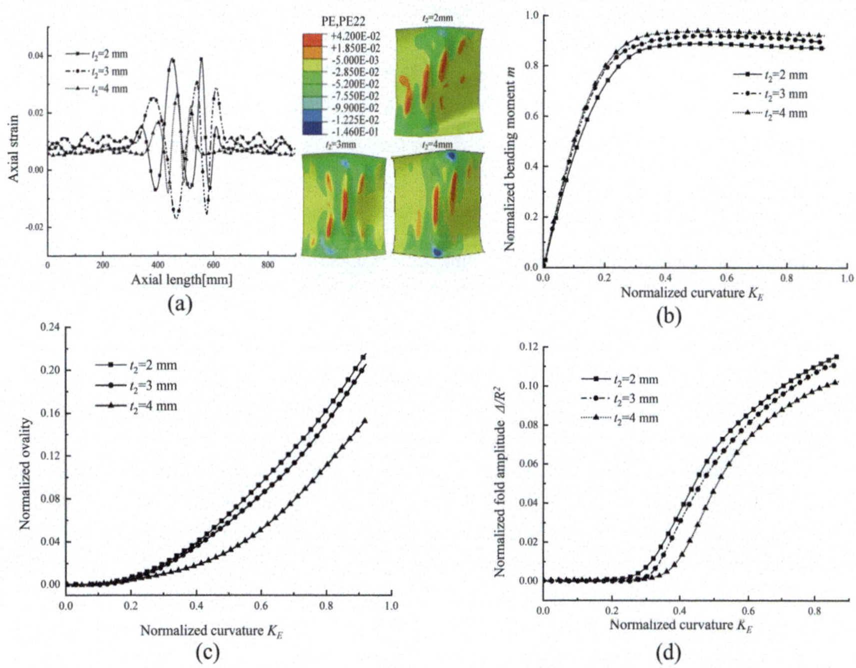

Fig. 6.13 Mechanical response of clad pipe with different wall thicknesses of lining pipes. **a** Plastic strain and its distribution of the liner pipe, **b** relationship between bending moment and curvature, **c** relationship between ovality and curvature, **d** relationship between fold amplitude and curvature

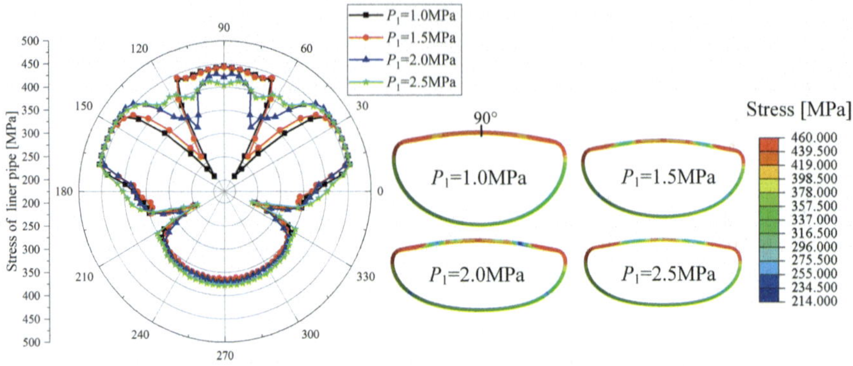

Fig. 6.14 Stress distribution of liner pipe section under different wall thicknesses

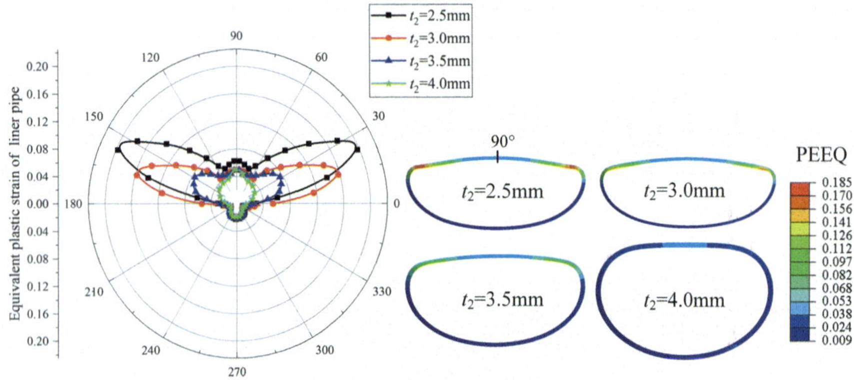

Fig. 6.15 Plastic strain distribution of liner pipe section under different wall thicknesses

Figure 6.16a shows the clad pipe's bending moment, initially increasing linearly (elastic deformation), then slowing down, peaking (plastic deformation), and finally decreasing (buckling), with higher wall thickness resulting in a higher maximum bending moment. In Fig. 6.16b, ovality increases with normalized curvature, indicating weakening buckling resistance, but less so with thicker walls. Figure 6.16c shows fold amplitude constant until KE = 0.6, then sharply increasing at KE = 0.8 (critical curvature), with thicker walls delaying wrinkling and reducing fold amplitude, indicating stronger buckling resistance.

6.4.2.4 Outer Pipe Wall Thickness

Figure 6.17 shows the mechanical behavior of composite pipes with varying wall thicknesses of the outer pipe. In Fig. 6.17a, an increase in the outer pipe's wall thickness leads to a decrease in the amplitude of plastic deformation, indicating a

Fig. 6.16 Mechanical response under different wall thicknesses of outer pipe. **a** Normalized bending moment, **b** normalized ovality, **c** normalized fold amplitude

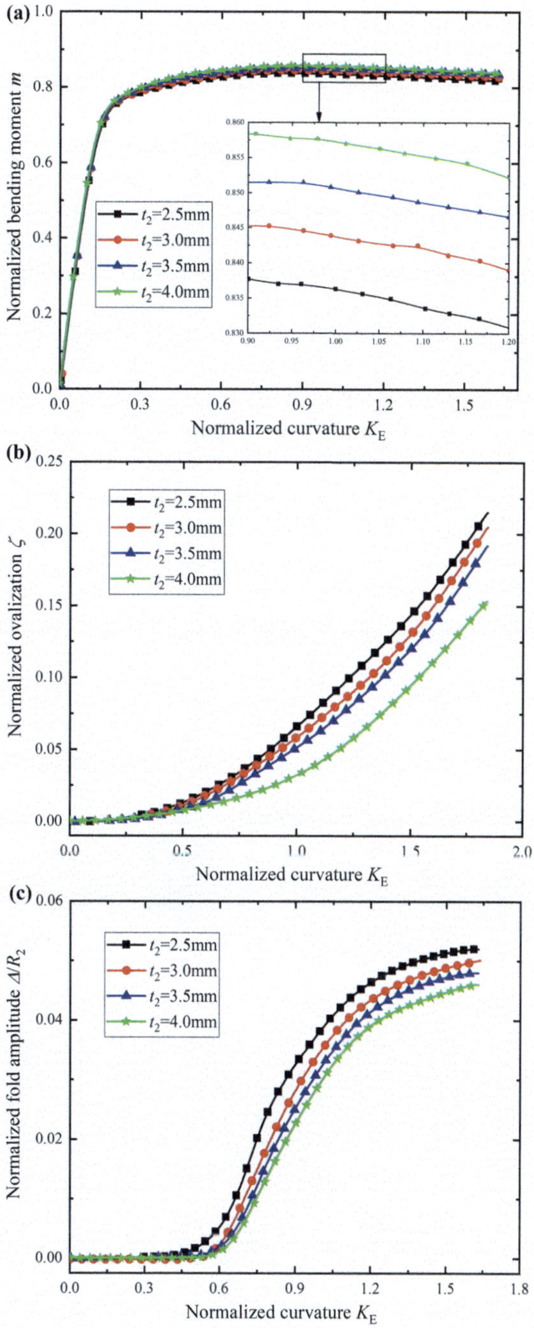

decrease in both plastic deformation and the number of wrinkles in the liner pipe. Figure 6.17b illustrates that a thicker wall of the outer pipe results in a higher bending moment for the composite pipe under identical curvature conditions, indicating an augmented resistance to bending moments. In Fig. 6.17c, a correlation is observed where the increased wall thickness of the outer pipe at a constant curvature yields a lower ovality, indicating an improvement in the buckling resistance of the liner pipe. Lastly, Fig. 6.17d shows that at a consistent curvature, an increase in the outer pipe's wall thickness corresponds to a decrease in the normalized fold amplitude of the liner pipe, further affirming its enhanced resistance to buckling.

Figure 6.18 displays the stress distribution in a liner pipe section affected by varying outer pipe wall thicknesses (t1). For $t_1 = 10$ mm and $t_1 = 15$ mm, maximum stress is found at 15°–40° and 140°–175°, located at the junction between compression and tension surfaces. For thicker walls, $t_1 = 20$ mm, and $t_1 = 25$ mm, maximum stress spans 15°–175° and − 180° to 0°, covering both compression and tension surfaces. As outer wall thickness increases, liner pipe stress decreases, indicating a shift of bending moment burden to the outer pipe, enhancing buckling prevention.

Figure 6.19a shows that increased outer wall thickness results in higher bending moments at the same normalized curvature, improving clad pipe resistance. In Fig. 6.19b, ovality grows slowly with normalized curvature, but for $t_1 \leq 15$ mm, buckling resistance is lost past $K_E = 1.2$. Thicker outer walls result in lower ovality

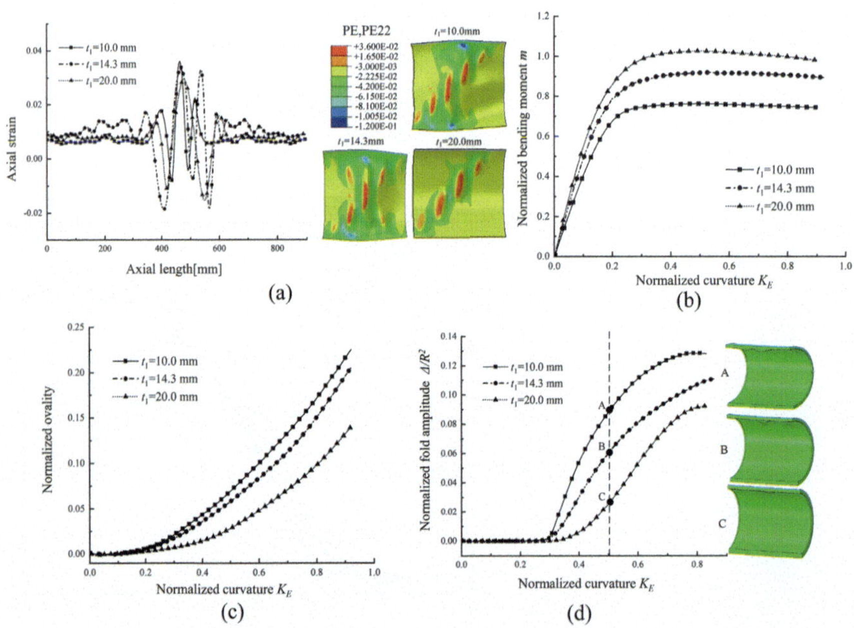

Fig. 6.17 Mechanical response of clad pipe with different wall thicknesses of outer pipes. **a** Plastic strain and its distribution of the liner pipe, **b** relationship between bending moment and curvature, **c** relationship between ovality and curvature, **d** relationship between fold amplitude and curvature

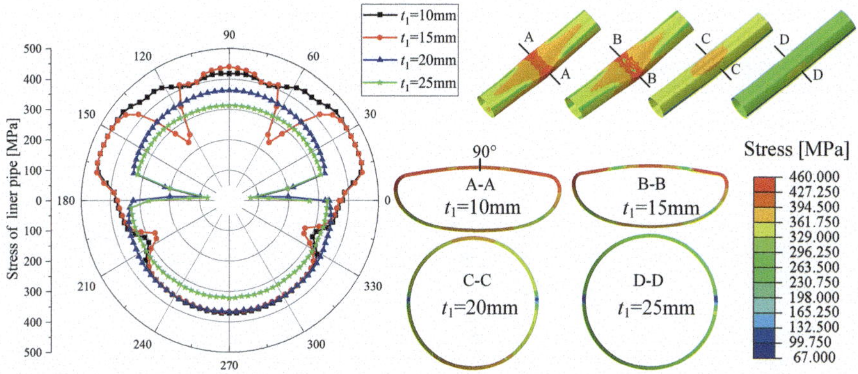

Fig. 6.18 Stress distribution curve of the liner pipe under different wall thicknesses of the outer pipe

at the same curvature, indicating improved bending moment resistance. Figure 6.19c shows constant fold amplitude until $K_E = 0.6$, with a critical curvature increase at $K_E = 0.7$ for $t_1 \leq 15$ mm, indicating buckling. For $t_1 > 15$ mm, the clad pipe shows better buckling resistance. Increased outer wall thickness results in lower normalized fold amplitude, showcasing stronger buckling resistance in the clad pipe.

6.5 Conclusions

Forming Pressure and Residual Stress: Optimal forming pressure is crucial to prevent the incomplete formation or buckling of the liner pipe, influencing the residual stress. Higher forming pressures increase residual stress, enhancing the pipe's ability to withstand bending moments and improving reliability.

Internal Working Pressure: Higher internal working pressures increase the bending moment while reducing ovality and fold amplitude, and enhance the liner pipe's buckling resistance, ensuring structural integrity.

Wall Thickness: Thicker walls in both the outer pipe and inner liner significantly improve bending and buckling resistance, crucial for performance and safety.

Buckling Behavior and Stress Distribution: Understanding how internal pressure, wall thickness, and material properties affect stress distribution and buckling behavior is vital for optimal pipe design.

Mechanical Response: The clad pipe's bending moment, ovality, and fold amplitude are influenced by internal pressure and wall thicknesses, with proper optimization leading to improved performance and buckling resistance.

These insights are essential for designing safer and more reliable bimetal mechanical clad pipes, ensuring they meet performance standards and safety margins.

Fig. 6.19 Mechanical response under different wall thicknesses of outer pipe. **a** Normalized bending moment, **b** normalized ovality, **c** normalized fold amplitude

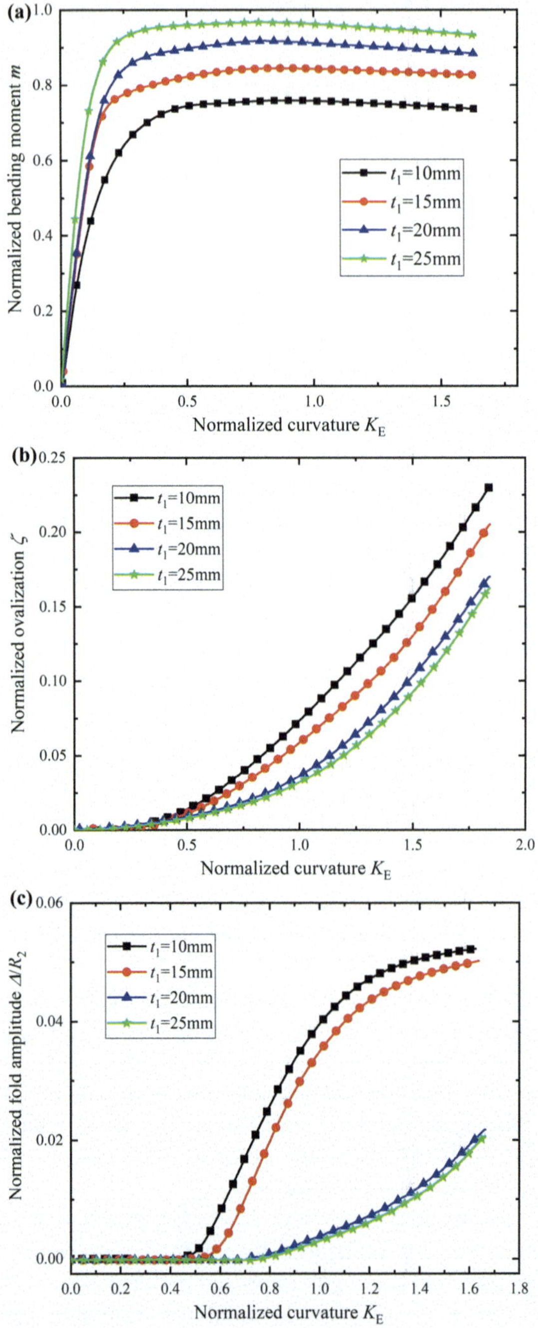

This chapter is adapted from the article ' Buckling Failure Mechanism of Liner Pipe in Bimetallic Mechanical Clad Pipe under Complex Loading,' originally published in *Ocean Engineering* (2022). DOI: https://doi.org/https://doi.org/10.1016/j.oceaneng.2022.112091. Reproduced with permission from Elsevier.

References

1. Alaswad A (2013) Computer aided design and optimization of bi-layered tube hydroforming process. Dissertation. Dublin City University
2. Vasilikis D, Karamanos SA (2012) Mechanical behavior and wrinkling of lined pipes. Int J Solids Struct 49(23–24):3432–3446. https://doi.org/10.1016/j.ijsolstr.2012.07.023
3. Dezhi Z et al (2014) Theoretical and experimental study of bimetal-pipe hydroforming. J Pressure Vessel Technol 136(6):061402. https://doi.org/10.1115/1.4026976
4. Yuan L, Kyriakides S (2014) Liner wrinkling and collapse of bi-material pipe under bending. Int J Solids Struct 51(3–4):599–611. https://doi.org/10.1016/j.ijsolstr.2013.10.026
5. Wang XS et al (2002) Calculation of hydro-forming pressure for bi-metal clad pipe. J Mech Strength 24(3):439–442

Chapter 7
Failure Mechanisms of Glass Fiber-Wound Composite Pipes

7.1 Introduction

This chapter delves into the failure mechanisms of fiber-wound composite pipes, crucial for various industries due to their high strength-to-weight ratio and corrosion resistance. Constructed by winding fibers, typically glass or carbon, around a mandrel and setting them with a resin matrix, common types include Glass Fiber Reinforced Polymer (GFRP) and Carbon Fiber Reinforced Polymer (CFRP) pipes [1]. Their increasing importance is attributed to superior mechanical properties and resistance to harsh conditions, necessitating a thorough understanding of their failure mechanisms for reliability and safety [2].

Failures can occur due to internal pressure, bending, torsion, and axial tension, making it imperative to investigate these mechanisms to optimize design and enhance performance. The first major topic explores the impact of winding angles on stress distribution and failure susceptibility in glass fiber-wound composite pipes under various loads, utilizing the three-dimensional Tsai-Wu failure criterion for analysis [3]. This provides insights for improved pipe design and material selection.

The second major topic focuses on nonmetallic composite pipes in hydrogen transportation, particularly basalt fiber materials. Simulation models under plate and spherical indenter loads examine the effects of structural parameters on mechanical behavior and failure modes, highlighting the weakest parts of the composite pipe under different conditions and guiding design optimization for safe and efficient hydrogen transportation.

Together, these investigations offer a comprehensive understanding of failure mechanisms in fiber-wound composite pipes, contributing to enhanced performance, reliability, and safety in their applications.

H. Zhang et al., *Submarine Pipeline Integrity: Assessment of Failure Modes and Advanced Evaluation Techniques*, Ocean Engineering & Oceanography 28,
https://doi.org/10.1007/978-3-031-92092-9_7

7.2 Failure Criteria

Hashin theory is a commonly used damage criterion for fiber reinforced composites. It considers four different damage initiation mechanisms: fiber tensile, fiber compression, matrix tensile and matrix compression [4]. The output variables associated with the four damage initiation criteria are used to indicate whether the criteria are met. If the value is less than 1, the damage initiation criterion is not met. If the value is greater than or equal to 1, the damage initiation criterion is satisfied. The initial conditions of damage are shown in Table 7.1.

The failure criterion of fiber wound composite pipe can be characterized by fracture strength of materials under different loads. When the stress in the fiber-wound layer reaches a certain threshold, failure initiates. Otherwise, there will be no damage. The failure of fiber-wound structures is a progressive process involving the gradual degradation of strength parameters. When the load reaches a certain threshold, the damage of fiber wound layer gradually begins. With the constant increasing of stress, the damage of fiber wound layer gradually expands until the stress reaches the bearing limit of fiber matrix material. Different failure types correspond to different strength criteria.

The three-dimensional Tsai-Wu criterion, a comprehensive stress interaction model, serves as a basis for evaluating the reinforced layer failure in fiber-wound composite pipes. This criterion posits that a composite ply is at risk of strength failure when the failure factor 'f' exceeds 1 and is considered safe when 'f' is below 1. A higher 'f' value under the same load indicates a higher likelihood of ply failure and a reduced load-bearing capacity of the composite pipe. For instance, in the presence of internal pressure, the reinforced layer with the highest 'f' is deemed the most vulnerable. By comparing these 'f' values, one can deduce the relative strength and load-bearing capacities of composite pipes with varying winding angles. The first layer failure theory, which asserts that the failure of one layer results in the failure of the entire composite layup, focuses on identifying these vulnerable layers. However, achieving a load condition where no composite pipe of any winding angle fails is challenging and requires numerous attempts. For simplicity, this study does not account for the stiffness degradation in the composite material post-failure, assuming that the

Table 7.1 Hashin failure criterion

Failure model	Failure criterion
Fiber tensile failure $\sigma_{11} > 0$	$\left(\frac{\sigma_{11}}{X^T}\right)^2 + \left(\frac{\tau_{12}}{S}\right)^2 \geq 1$
Fiber compression failure $\sigma_{11} < 0$	$\left(\frac{\sigma_{11}}{X^C}\right)^2 \geq 1$
Matrix tensile failure $\sigma_{22} > 0$	$\left(\frac{\sigma_{22}}{Y^T}\right)^2 + \left(\frac{\tau_{12}}{S^L}\right)^2 \geq 1$
Matrix compression failure $\sigma_{22} < 0$	$\left(\frac{\sigma_{22}}{2S^{Tr}}\right)^2 + \left[\left(\frac{Y^C}{2S^{Tr}}\right)^2 - 1\right]\frac{\sigma_{22}}{Y^C} + \left(\frac{\tau_{12}}{S^L}\right)^2 \geq 1$

reinforced layer retains its original mechanical properties even when 'f' surpasses one. The failure factor 'f' from the three-dimensional Tsai-Wu criterion is derived using isotropic stresses in the composite layup.

$$\begin{aligned} f = {} & F_1 S_{11} + F_2 S_{22} + F_3 S_{33} + F_{11} S_{11}^2 \\ & + F_{22} S_{22}^2 + F_{33} S_{33}^2 + 2F_{12} S_{11} S_{22} + 2F_{23} S_{22} S_{33} \\ & + F_{44} S_{23}^2 + F_{55} S_{13}^2 + F_{66} S_{12}^2 \end{aligned} \tag{7.1}$$

where,

$$\begin{cases} F_1 = \frac{1}{X_T} - \frac{1}{X_C}, \; F_{11} = \frac{1}{X_T X_C}, \; F_1 = \frac{1}{S_{23}^2} \\ F_2 = \frac{1}{Y_T} - \frac{1}{Y_C}, \; F_{22} = \frac{1}{Y_T Y_C}, \; F_1 = \frac{1}{S_{13}^2} \\ F_1 = \frac{1}{Z_T} - \frac{1}{Z_T}, \; F_{33} = \frac{1}{Z_T Z_C}, \; F_1 = \frac{1}{S_{12}^2} \end{cases} \tag{7.2}$$

According to Ref. [3], when the material is transversely isotropic, $F_{12} = F_{13} = -\frac{1}{2}\sqrt{F_{11}F_{22}}$, $F_{23} = F_{22} - \frac{1}{2}F_{44}$. Where, $S_{11}, S_{22}, S_{33}, S_{12}, S_{13}, S_{23}$ are the composite layup isotropic stresses and $X_T, X_C, Y_T, Y_C, Z_T, Z_C, S_{12}, S_{13}, S_{23}$ are the strength parameters of glass fibers.

7.3 Numerical Simulation Model

The fiber-wound composite pipe, illustrated in Fig. 7.1, comprises a three-layer structure: an inner liner, a reinforced layer, and an outer cover, bonded together through a hot-melt process. The inner liner and outer cover, typically made of corrosion-resistant thermoplastics like HDPE or PP, protect against fluid penetration. The reinforced layer, serving as the main load-bearing structure, consists of carbon, glass, or aramid fiber tapes wound at specific alternating angles.

In this model, HDPE constitutes the inner liner and outer cover, while glass fiber unidirectional tape reinforces the structure, with material parameters based on reference [5]. The composite pipe features an inner diameter of 48 mm, an outer diameter of 69 mm, and 4 mm-thick inner and outer layers. The reinforcement comprises 10 layers, each 0.25 mm thick, totaling 2.5 mm, with a fiber winding angle of ± 55°. To minimize boundary condition effects, the pipe's length adheres to St. Venant's principle, being at least five times its diameter [6], and stress analysis focuses on the pipe body's center [7]. Table 7.2 details the working conditions' loads.

The typical load forms are spherical indenter load and plate load, and the models are shown in Fig. 7.2. Under spherical indenter loading conditions, pipe diameter is 416 mm, length is 2000 mm and thickness is 4.8 mm. The layup scheme follows $[45/0/-45/90]_6$, and a spherical indenter with a diameter of 200 mm is applied. The model of continuous shell element exhibits a more stable response under compressive load. It is used to simulate the pipe in both models, while discrete elements are used for

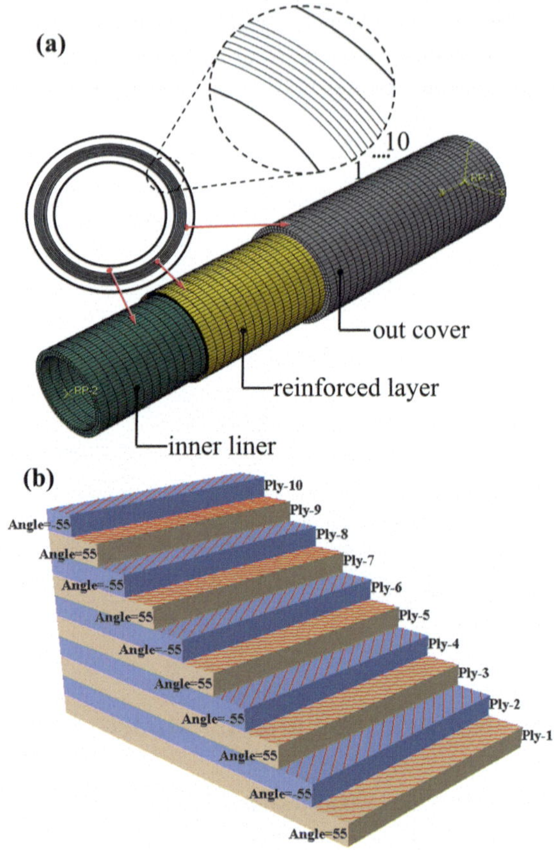

Fig. 7.1 Structure of fiber-wound composite pip. **a** Finite element model, **b** winding Angle ± 55° reinforced lay

Table 7.2 Load conditions

Working conditions	Load
Internal pressure	20 MPa internal pressure, 36.191 kN axial tension
Torsion	2 kN m
Axial tensile	72 kN
Bending load	1 kN m

the rigid plate and indenter. The parameters of basalt fiber materials used are shown in Table 7.3.

Yao et al. [8] performed experiments to measure the torsional stiffness of fiber-wound composite pipes. A numerical model, mirroring the parameters of the experimental sample, was developed and subjected to a rotation angle of 0.4 rad. As depicted in Fig. 7.3, the model's torsional stiffness closely aligns with the experimental findings. The simulation yields a torsional stiffness of 8229.95 N m^2/rad, compared to the experimental result of 8333 N m^2/rad from reference [8], resulting

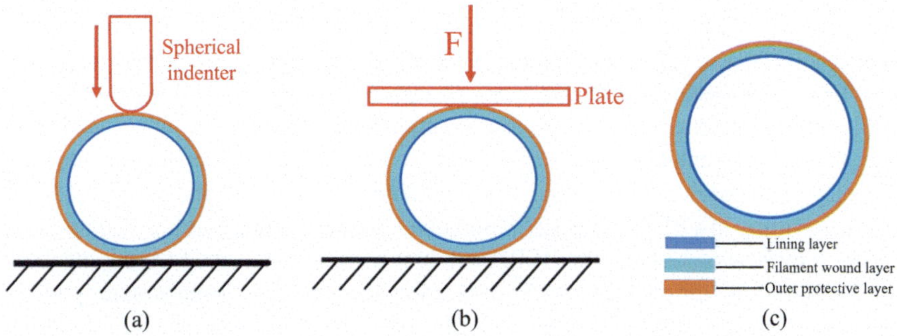

Fig. 7.2 Schematic diagram of the calculation model. **a** Spherical loading model, **b** plate loading model, **c** composite pipe model

Table 7.3 Parameters of basalt fiber composites

E_X/MPa	E_Y/MPa	E_Z/MPa	V_{XY}/ MPa	V_{YZ}/ MPa	V_{XZ}/ MPa	G_{XY}/ MPa	G_{YZ}/ MPa	G_{XZ}/ MPa
65,000	7600	7600	0.28	0.047	0.28	4840	4980	4840
X_T/MPa	X_C/MPa	Y_T/MPa	Y_C/ MPa	Z_T/MPa	Z_C/ MPa	S_{XY}/ MPa	S_{XZ}/ MPa	S_{YZ}/ MPa
1200	850	74	180	74	180	50	50	50

in a minor error of 1.24%. This demonstrates the simulation method's reliability, ensuring confidence in its application for further analysis.

Figure 7.4 displays the maximum load capacities of two models with varying mesh densities. A stable trend in maximum load alongside a consistent displacement

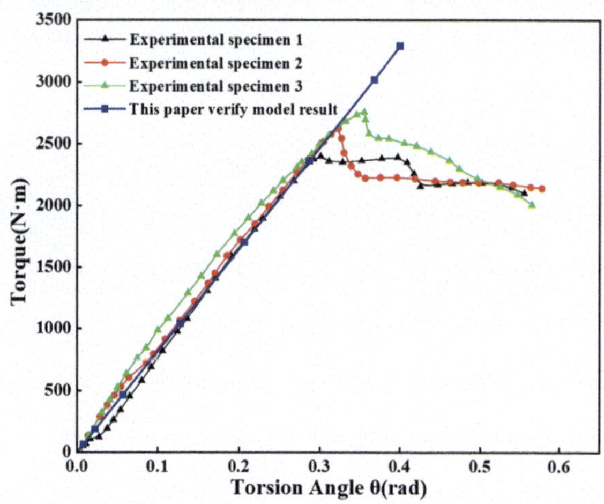

Fig. 7.3 Torque-torsion angle curves of simulation and experimental

change relationship indicates good model convergence. For precise results in subsequent analyses, the spherical indenter load model utilized 26,400 grids, while the plate load model used 52,800 grids.

Figure 7.5 presents force–displacement curves for a composite pipe under plate load, comparing simulation and experimental results. The test specimen, consisting of six layers of basalt fiber prepreg, was wound onto a core mold and cured at 180 °C for 3 h, resulting in a BFRP with a 27.9 mm inner diameter and 1.8 mm wall thickness. Radial compression tests on the BFRP, conducted using a universal testing machine at a 2 mm/min loading speed and 30 mm radial compression displacement, yielded a maximum load of 946 N experimentally and 931 N via simulation. The small discrepancy confirms the numerical model's reliability.

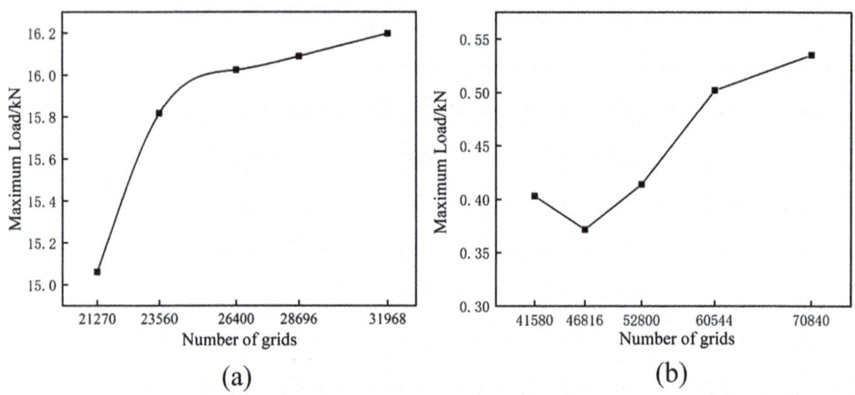

Fig. 7.4 Mesh independence verification. **a** Spherical model, **b** Plate model

Fig. 7.5 Comparison between experiment and simulation results

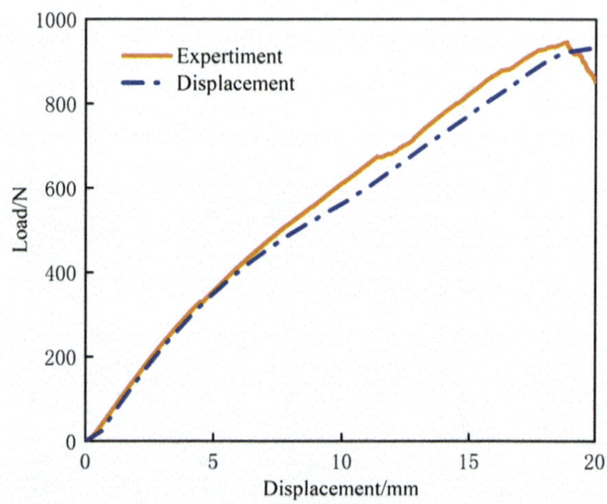

7.4 Results and Discussions

7.4.1 Failure Mechanisms of the Reinforced Layer of Glass Fiber-Wound Composite Pipe

The winding angle of the reinforcement layer is one of the most important structural parameters of the composite pipe. It is of great significance to study its influence on the mechanical response and bearing capacity of composite pipe. Therefore, the winding angle of $\pm$ 20° ~ $\pm$ 80° is selected for analysis.

7.4.1.1 Internal Pressure Load

The composite pipe is subjected to both circumferential and axial biaxial stresses during the internal pressure test [9]. Therefore, the internal pressure load analysis needs to consider the end force, where end force $F = P\pi R^2$, R is the internal diameter and P is the internal pressure load.

Figure 7.6 illustrates the stress distribution of the fiber-wound composite pipe subjected to an internal pressure of 20 MPa with a winding angle of $\pm$ 55°. The maximum stresses in the inner liner and outer cover are 21.1 MPa and 9.488 MPa, respectively. In contrast, the reinforced layer experiences a maximum stress of 230.8 MPa, which is significantly higher than those of the inner liner and outer cover. This indicates that the reinforced layer is the primary load-bearing component of the fiber-wound composite pipe.

Figure 7.7 illustrates the anisotropic stress distribution in the reinforced layer at a winding angle of $\pm$ 55°. Examining the laminate from the inner to the outer layers, the longitudinal fiber stress (S11) gradually decreases. In contrast, the transverse stress (S22) perpendicular to the fiber direction increases progressively. The interlaminar stress (S33) is compressive and decreases gradually. The fiber/matrix shear stress (S12) alternates between positive and negative values, with its magnitude diminishing over the layers. The shear stresses S13 and S23 are negligible, approaching zero. These results are consistent with the stress variation behavior of fiber-wound composite pipes reported in Ref. [5].

At an internal pressure of 20 MPa, the stresses in the reinforced layer of the composite pipe for various winding angles are shown in Fig. 7.8. The fiber longitudinal and transverse stresses are tensile, while the interlaminar stresses are compressive. For winding angles between $\pm$ 20° and $\pm$ 80°, both the fiber longitudinal stress and interlaminar stress decrease gradually from the inner layers to the outer layers. The transverse stress decreases from inside to outside when the winding angle is less than $\pm$ 45°, and increases when the winding angle exceeds $\pm$ 45°. Due to adjacent fiber layers being wound at opposite angles, the fiber-matrix shear stress alternates in sign from the inner to outer layers: it changes from positive to negative for winding angles less than $\pm$ 65°, and from negative to positive for angles greater than $\pm$ 65°. Therefore, one cannot consider the innermost layer as the most failure-prone based

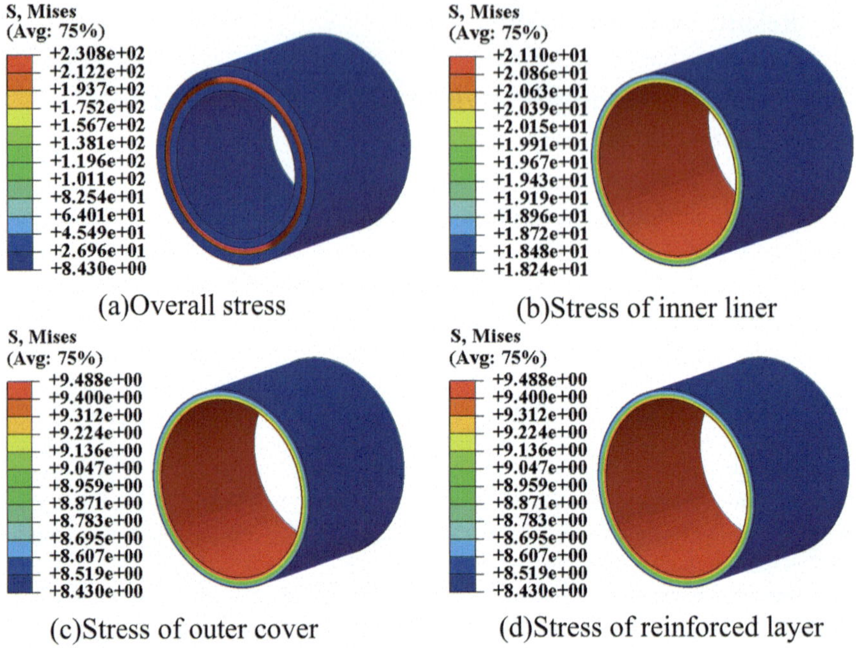

Fig. 7.6 Stress distribution in each part of the composite pipes. **a** Overall stress, **b** stress of inner liner, **c** stress of outer cover, **d** stress of reinforced layer

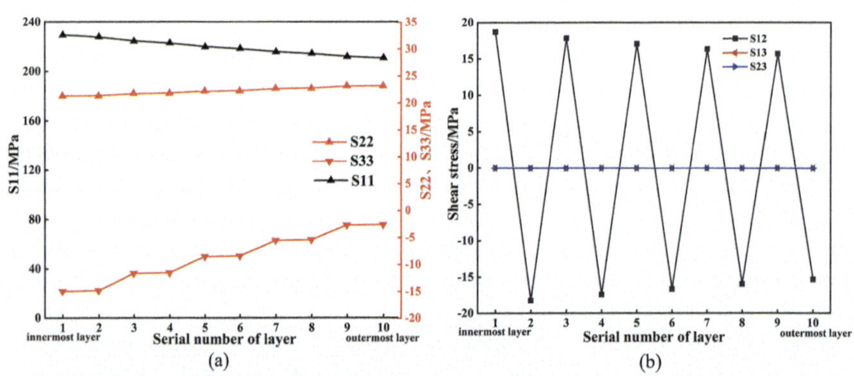

Fig. 7.7 Stress distribution of reinforced layer at winding angle $\pm$ 55°. **a** Fiber longitudinal, transverse and interlaminar stress, **b** shear stress

solely on individual stress components; a comprehensive evaluation of stresses in all directions using the Tsai-Wu criterion is necessary.

Figure 7.9a displays the failure factors for each layer of the reinforced composite pipe at different winding angles. The failure factors decrease from the inner to the outer layers, indicating that the innermost layer is most prone to failure.

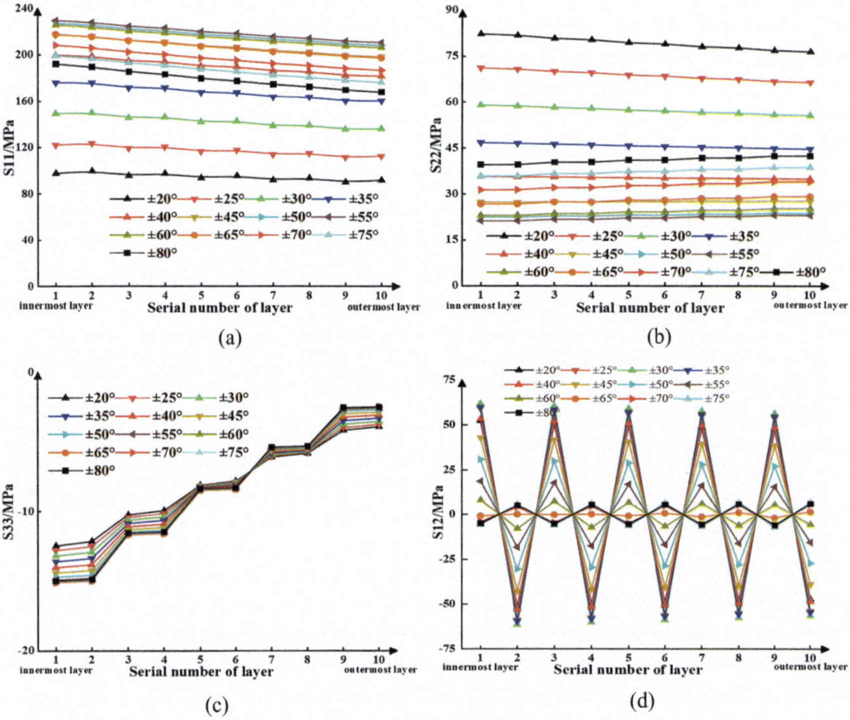

Fig. 7.8 Stress of reinforced layer under internal pressure load. **a** Longitudinal stress of fiber, **b** transverse stress, **c** interlaminar stress, **d** fiber/matrix shear stress

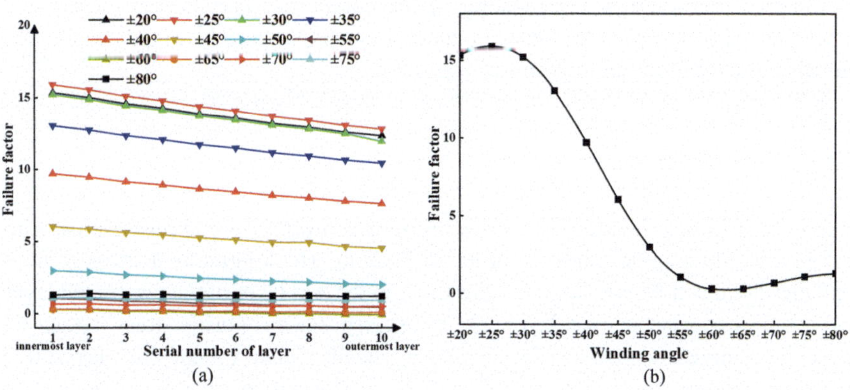

Fig. 7.9 Failure factor of reinforced layer under internal pressure load. **a** The failure factor of each layer, **b** the failure factor of innermost layer

Figure 7.9b illustrates how the failure factor of the innermost layer varies with winding angle under internal pressure loading. The failure factor increases slightly from ± 20° to ± 25°, decreases rapidly from ± 25° to ± 55°, and then decreases slightly after ± 55°, reaching a minimum around ± 60°.

The composite material's transverse and shear strengths are significantly lower than the fibers' longitudinal strength, while transverse and shear stress values are relatively high in composite pipes. Consequently, transverse and shear stresses are the primary factors contributing to failure. Beyond a winding angle of ± 60°, the fiber transverse (matrix) component becomes the main load-bearing element, leading to a significant increase in stress and a minor increase in shear stress, causing the failure factor to rise slightly. Maximum transverse and shear stresses occur at a winding angle of approximately ± 25°, resulting in the peak failure factor; thus, ± 25° is the least favorable for carrying internal pressure.

For winding angles exceeding ± 55°, failure factors tend to stabilize at lower values compared to other angles. Therefore, considering both the difficulty of winding molding and load-bearing capacity, a winding angle of ± 55° is deemed optimal for carrying internal pressure.

Reference [10] mentions that when the pipe is subjected to internal pressure and its end forces generated by the internal pressure, the ratio of annular stress to axial stress is 2:1. Therefore, the optimum winding angle derived here is in accordance with the previously mentioned ± 55° as the optimum winding angle when the ratio of annular stress to axial stress is 2:1.

In terms of the overall variation of the failure factor with angle, the following assumptions can be made: The composite pipe with a small winding angle has low pressure-bearing capacity, and the large winding angle is conducive to improving the internal pressure-bearing capacity of the composite pipe, but an excessively large winding angle provides limited improvement in pressure-bearing capacity and increases the difficulty of winding and forming.

7.4.1.2 Torsional Load

Figure 7.10 shows the stresses in the reinforced layer of a composite pipe with different winding angles under a torque of 2 kN m. The longitudinal stress of fiber and interlaminar stress are compressive stress in the odd layer of the reinforced layer and tensile stress in the even layer. The transverse stress is tensile stress in the odd layer and compressive stress in the even layer. The fiber/matrix shear stress is positive when the winding angle is greater than ± 45° and negative when it is less than ± 45°. All four stresses increase gradually from the inner to the outer layers.

Figure 7.11a presents the failure factors for each layer of the reinforced composite pipe under torsional loading at various winding angles. The failure factors of odd-numbered layers are higher than those of even-numbered layers due to the composite material's high tensile strength and low compressive strength along the fiber direction. The increasing trend of failure factors from the inner to the outer layers across all winding angles indicates that the outer layer is more prone to failure. This observation

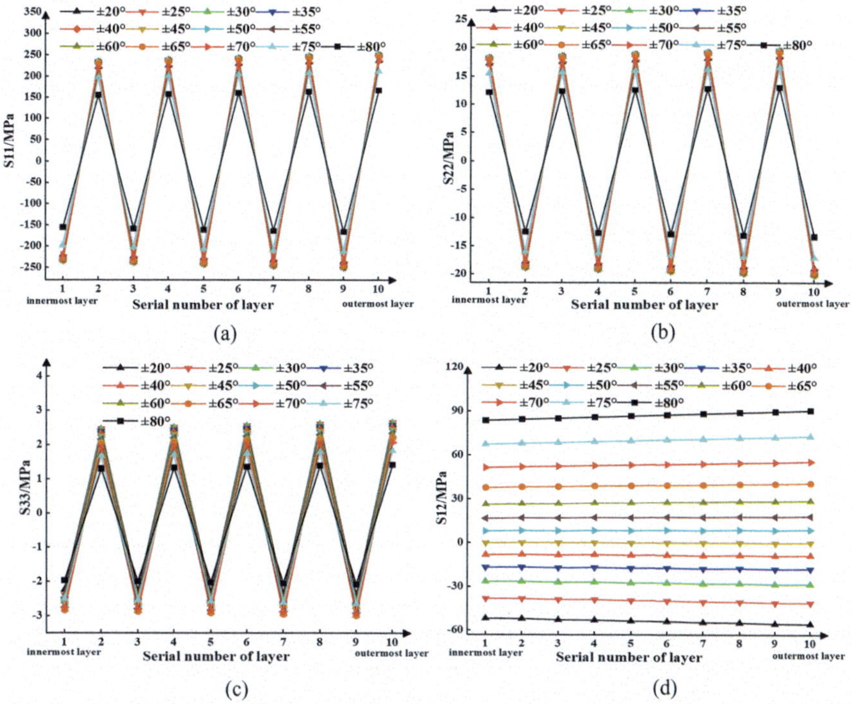

Fig. 7.10 Stress of reinforced layer under torsional load. **a** Longitudinal stress of fiber, **b** transverse stress, **c** interlaminar stress, **d** fiber/matrix shear stress

aligns with the failure locations observed in the torsional stiffness experiments of fiber-wound composite pipes by [7], thereby validating the research methodology used in this study.

Figure 7.11b illustrates the variation of the failure factors for the 9th and 10th layers with respect to winding angle under torsional loading. The failure factor curve exhibits a U-shaped pattern. Failure factors within the winding angle range of $\pm$ 35° to $\pm$ 55° are significantly lower than those in other ranges, indicating that the optimal winding angle interval for fiber-wound composite pipes under torsional load is between $\pm$ 35° and $\pm$ 55°.

7.4.1.3 Axial Tensile Load

Figure 7.12 presents the stress distribution across each layer of the reinforced composite pipe under a 72 kN axial tensile load at varying winding angles. As the winding angle increases, the longitudinal stress in the fiber transitions from tensile to compressive, while the transverse stress shifts from compressive to tensile. For winding angles between $\pm$ 60° and $\pm$ 70°, the longitudinal fiber stress decreases from the inner to outer layers, but increases at other winding angles. At winding angles

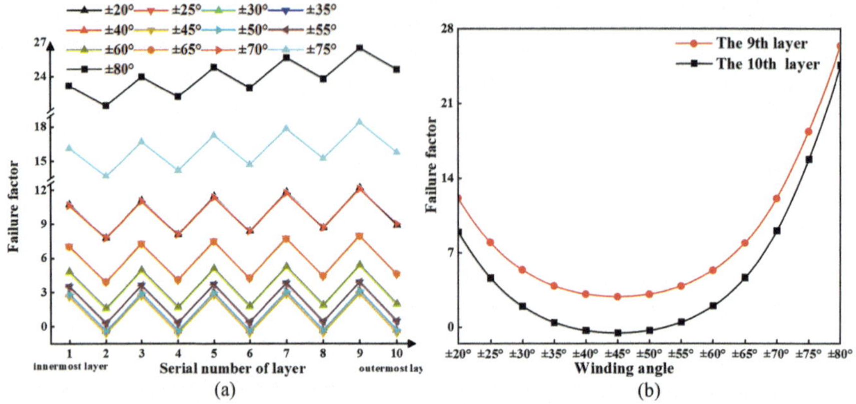

Fig. 7.11 Stress of reinforced layer under torsional load. **a** The failure factor of each layer, **b** the failure factor of the 9th and 10th

of ± 35° to ± 55°, the transverse stress increases outward, while at other angles, it decreases outward. The interlaminar stress decreases from the first to the seventh layer, showing compressive stress at angles below ± 50° and tensile stress above ± 55°. From the eighth to the tenth layer, the interlaminar stress generally increases, displaying tensile stress below ± 50° and compressive stress above ± 55°. The fiber/matrix shear stress alternates between negative and positive, with its absolute value remaining relatively constant.

Figure 7.13a shows the failure factors for each layer of the reinforced composite pipe under tensile load at various winding angles. For winding angles below ± 60°, the failure factor decreases from the inner to outer layers, while for angles above ± 60°, it increases outward. This indicates that the failure-prone layer shifts based on the winding angle: the inner layer is more vulnerable when the winding angle is less than ± 60°, and the outer layer becomes more failure-prone at angles greater than ± 60°.

Figure 7.13b illustrates the variation of the failure factor for the most critical layer of the reinforced composite pipe under tensile load. The failure factor increases slowly between ± 20° and ± 30°, rises rapidly between ± 30° and ± 60°, and then decreases sharply beyond ± 60°. The failure factor is significantly lower for winding angles in the ± 20° to ± 30° range, suggesting that smaller winding angles improve the tensile strength of fiber-wound composite pipes. This contrasts with the optimal winding angle range for internal pressure, indicating that a single winding angle may not be suitable for multiple load conditions. Therefore, a design approach for composite pipes could involve separate reinforced layers optimized for pressure and tensile loads.

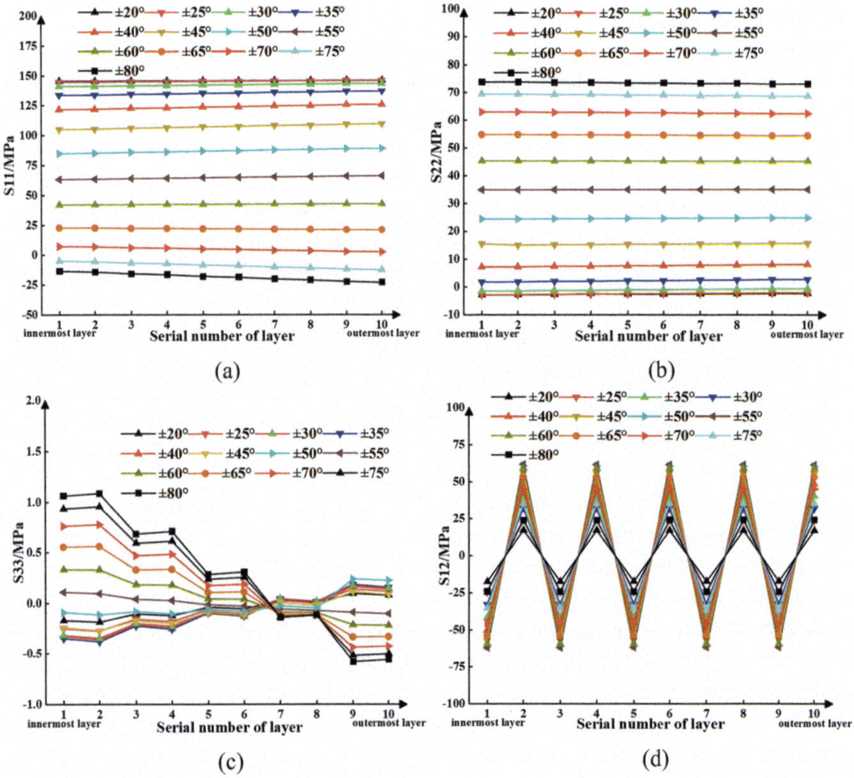

Fig. 7.12 Stress of reinforced layer under tensile load. **a** Longitudinal stress of fiber, **b** transverse stress, **c** interlaminar stress, **d** fiber/matrix shear stress

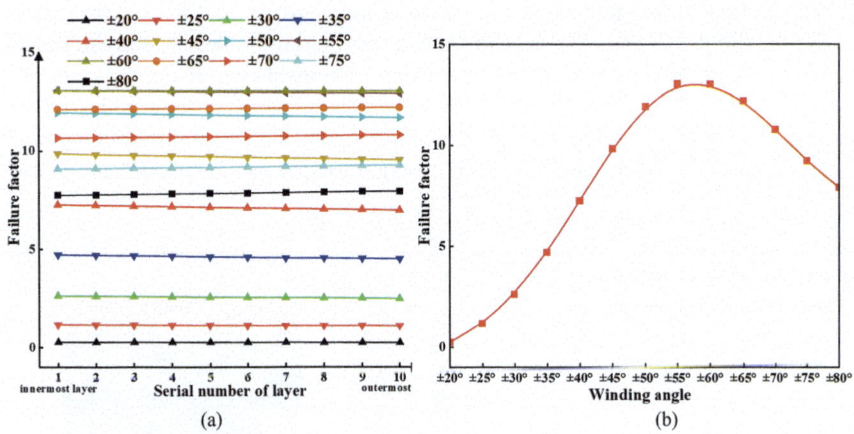

Fig. 7.13 Failure factor of reinforced layer under tensile load **a** The failure factor of each layer, **b** the failure factor of dangerous layer

7.4.1.4 Bending Load

As shown in Fig. 7.14, the bending load produces two high stress zones on both sides of the composite tube i.e. 'inside of bending' and 'outside of bending'.

Figure 7.15 is the stress of each layer inside and outside the bending of the reinforced layer of the composite pipe with different winding angles under the pure bending load of 1kN·m. The longitudinal stresses of fiber are equal in magnitude but opposite in direction on the inside and outside of the bending. As the winding angle increases, the longitudinal stress of fiber compressive stress changes to tensile stress on the inside of the bending, while on the outside of the bending it changes from tensile to compressive stress. The longitudinal stress of fiber gradually decreases from inside to outside at winding angles of ± 65° to ± 70°, and gradually increases from inside to outside at other winding angles.

The transverse stress on the inner and outer sides of the bending is equal in magnitude but opposite in direction. As the winding angle increases, the transverse stress on the inside of the bend shifts from tensile to compressive, and from compressive to tensile on the outside. For winding angles between ± 20° and ± 30°, the transverse stress decreases from inside to outside, while at other angles it increases outward.

The interlaminar stress follows a similar pattern, with equal magnitudes but opposite directions on the inner and outer sides of the bending. For winding angles between ± 20° and ± 50°, the interlaminar stress on the inside changes from tensile to compressive, and from compressive to tensile on the outside. For winding angles between ± 55° and ± 80°, this trend reverses. Overall, interlaminar stress decreases from inside to outside in layers 1 through 6, while it increases in layers 7 through 10.

The fiber/matrix shear stress also alternates in sign from inside to outside and follows the same "anti-symmetric" pattern as the other stresses, with equal magnitudes but opposite directions on the inner and outer sides. Under bending loads, the longitudinal stress, transverse stress, interlaminar stress, and fiber/matrix shear stress all exhibit this anti-symmetric behavior. Furthermore, there is consistency in the stress distribution patterns across the layers from inside to outside.

Figure 7.16a, b display the failure factors on the inner and outer sides of the reinforced composite pipe under a pure bending load of 1kN·m at different winding angles. The failure factor increases from the inner to the outer layers, indicating that

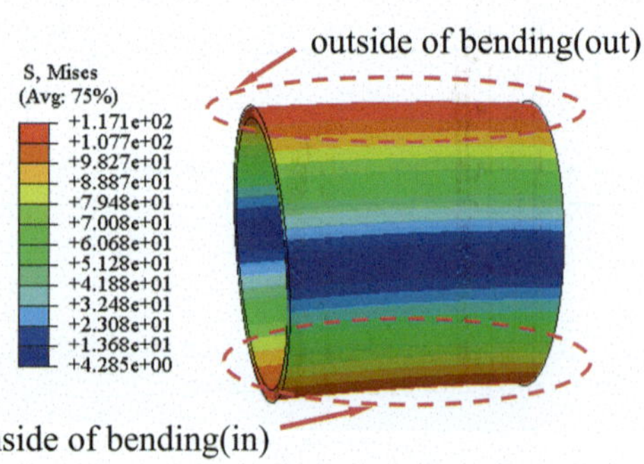

Fig. 7.14 Stress of reinforced layer at a winding angle of ± 55°under bending load

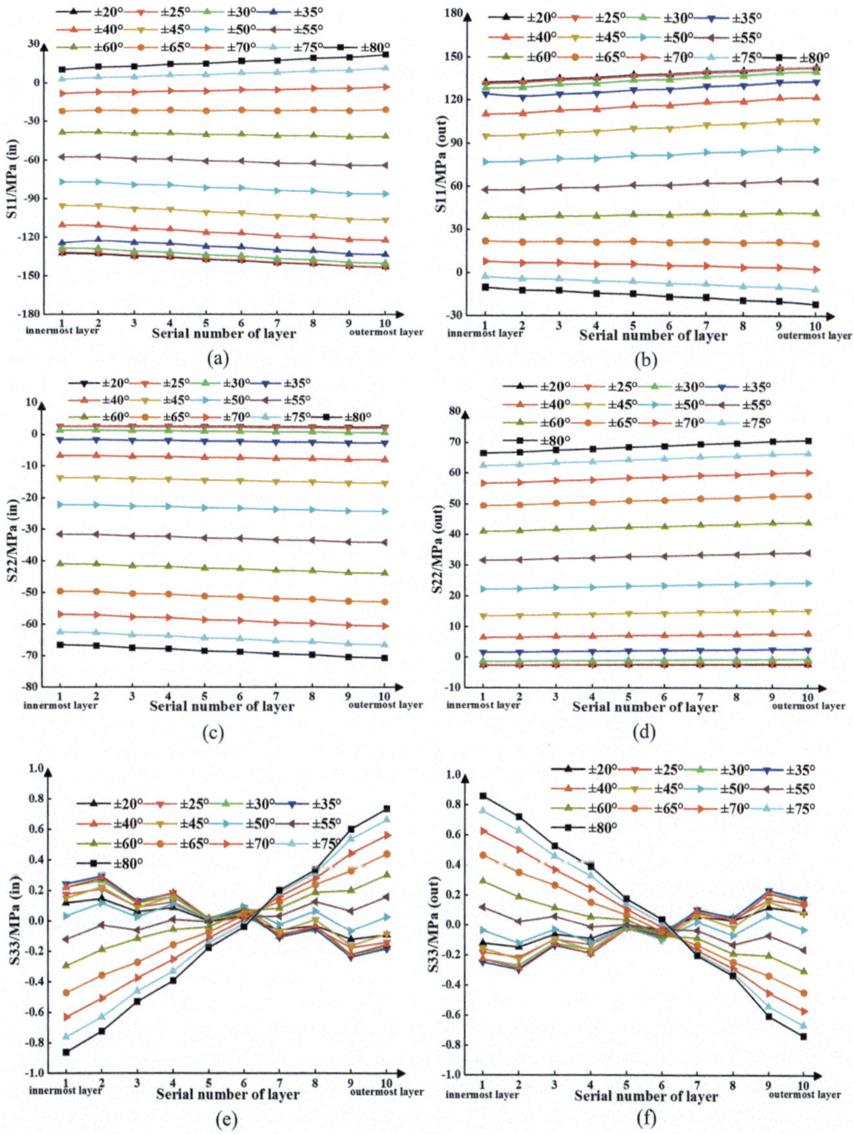

Fig. 7.15 Stress of reinforced layer under bending load. **a** Longitudinal stress of inside of bending, **b** longitudinal stress of outside of bending, **c** transverse stress of inside of bending, **d** transverse stress of outside of bending, **e** interlaminar stress of inside of bending, **f** interlaminar stress of outside of bending, **g** fiber/matrix shear stress of inside of bending, **h** fiber/matrix shear stress of outside of bending

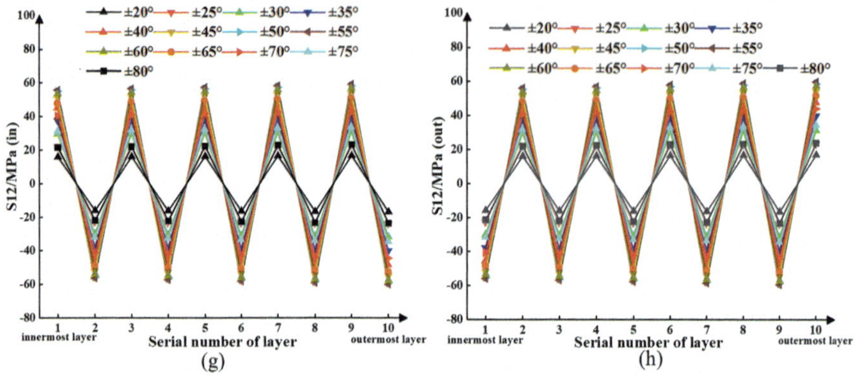

Fig. 7.15 (continued)

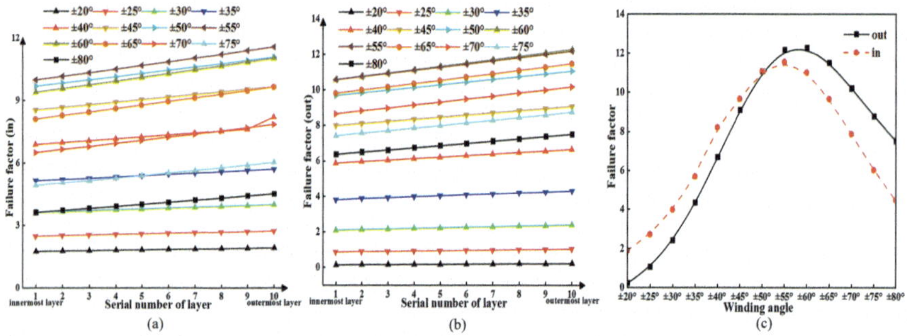

Fig. 7.16 Failure factor of reinforcement layer under bending load. **a** The failure factor of inside the bending, **b** the failure factor of outside the bending, **c** the failure factor of dangerous layer

the outer layer is more prone to failure and represents the hazardous layer under bending loads.

Figure 7.16c illustrates the variation of the hazardous layer failure factor with respect to winding angle. For angles between $\pm 20°$ and $\pm 50°$, the failure factor on the inner side of the bending is higher than on the outer side. Conversely, for winding angles between $\pm 50°$ and $\pm 80°$, the failure factor on the outer side exceeds that of the inner side. This suggests that for winding angles between $\pm 20°$ and $\pm 50°$, failure is more likely to occur on the inner side of the bend, while for angles between $\pm 50°$ and $\pm 80°$, failure is more likely on the outer side. The failure factor is lowest for both the inner and outer sides when the winding angle is between $\pm 20°$ and $\pm 30°$, indicating that smaller winding angles improve the bending resistance of fiber-wound composite pipes.

7.4.2 *Failure Mechanisms of the Reinforced Layer of Basalt Filament-Wound Composite Pipe*

7.4.2.1 Spherical Extrusion Condition

Stress Distribution: The stress in BFRP under a spherical indenter load is initially concentrated at the contact surface, spreading axially and circumferentially as the indenter displaces 6% of the pipe diameter. At 20% displacement, stress concentrates at the dented edge and the contact area with the spherical indenter (Fig. 7.17).

Force–Displacement Relationship: The force increases rapidly until K1, with the composite pipe resisting the indenter load through fiber elasticity. Between K1 and K2, internal damage occurs, slowing the force growth until it peaks. Post-K2, the pipe stiffness and force decrease sharply, indicating failure, until K3, where undamaged areas continue to resist deformation (Fig. 7.18).

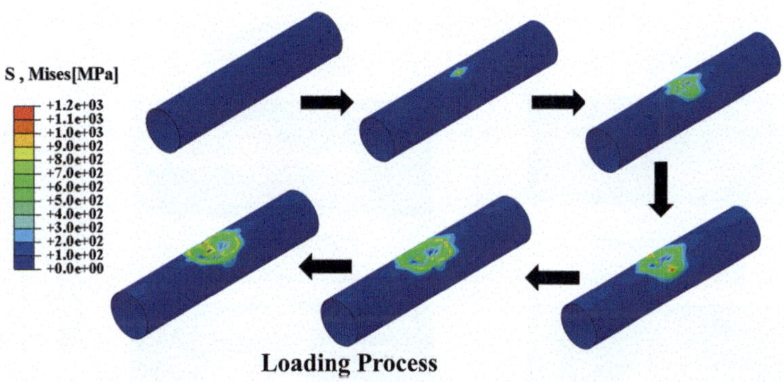

Fig. 7.17 Stress distributions of the pipe under spherical indenter load

Fig. 7.18 Force–displacement curve under spherical indenter load

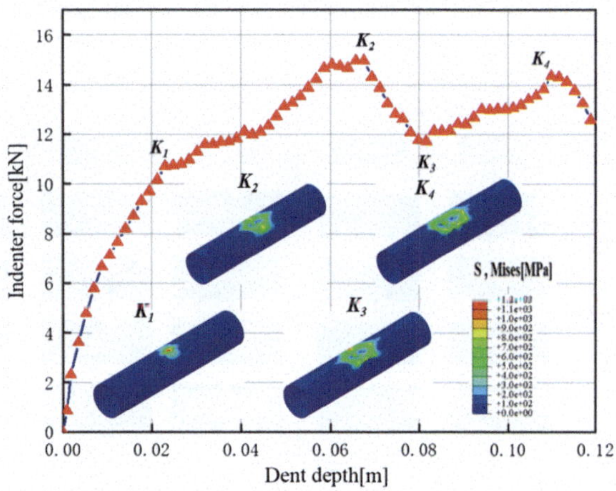

Layer-Specific Damage: Under spherical indenter load, Ply1-45° shows compression damage concentrated at the dent edge and tensile damage along the pipe's central axis and circumferential line. Matrix compression damage is more widespread, diffusing outward along the pipe's axis (Figs. 7.19 and 7.20).

Strain Distribution: Axial strains are predominantly compressive at the dent, with tensile strains appearing in line with the fiber direction in Ply21(45°) and Ply23(−45°). Circumferential strains follow a similar pattern, with variations across different fiber layers (Fig. 7.21a, b).

Influence of Diameter-Thickness Ratio

Load Response: The maximum depression depth of the composite pipe is directly proportional to the diameter-to-thickness ratios. Pipes with smaller ratios can withstand larger indenter forces without significant deformation, while larger ratios result in greater deformation but reduced ultimate loads (Figs. 7.22 and 7.23).

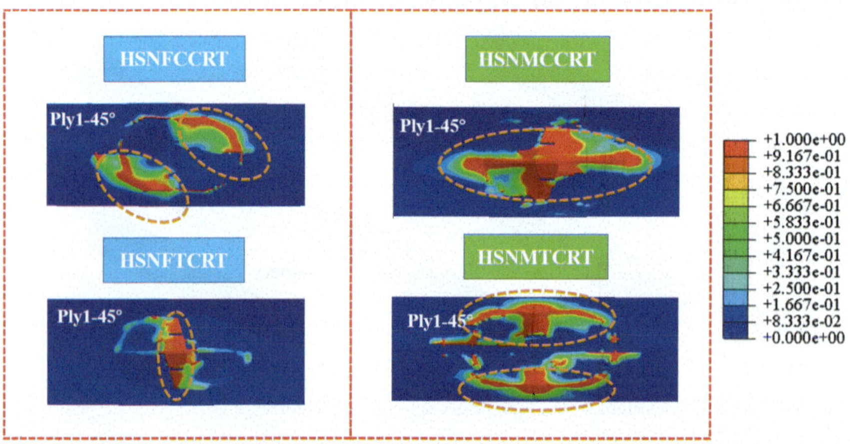

Fig. 7.19 Damage diagram of Ply1-45deg under spherical indenter load

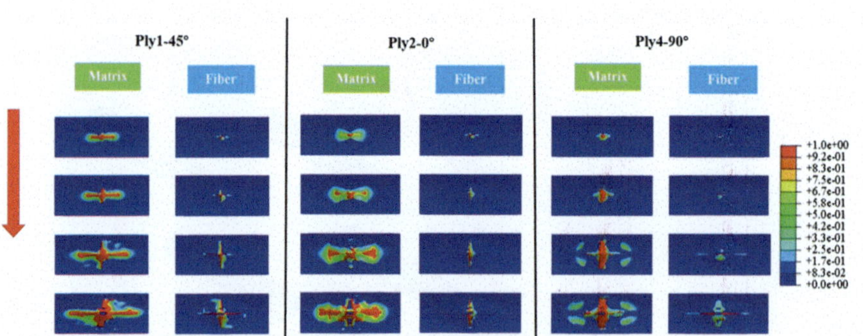

Fig. 7.20 Tensile damage change of each layer under spherical indenter load

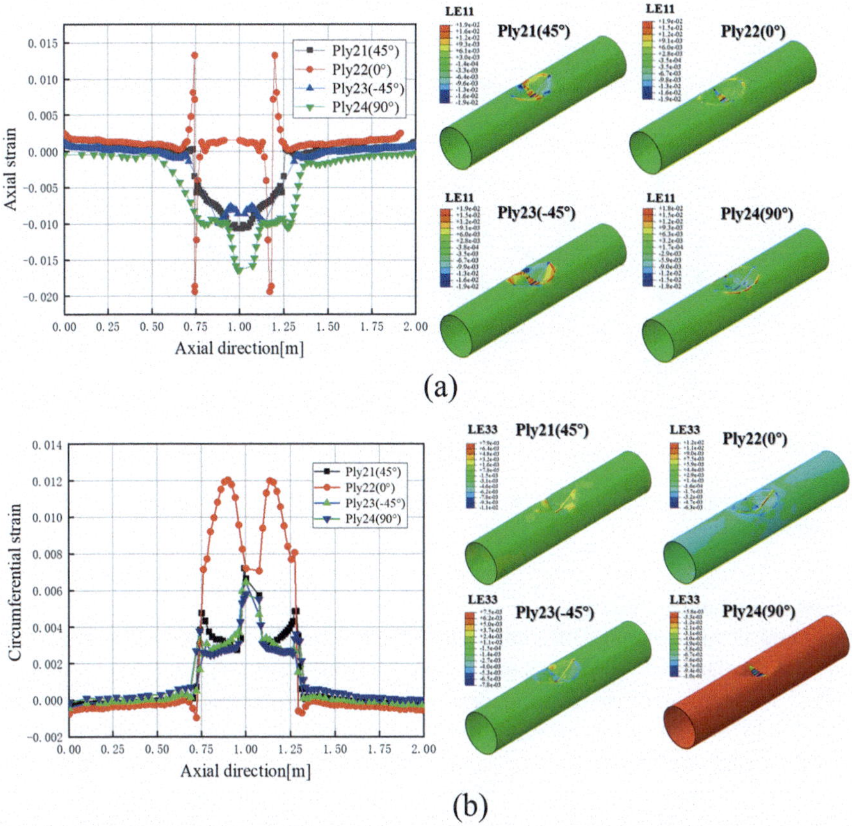

Fig. 7.21 Strain distribution of each layer under spherical indenter load. **a** Axial strain, **b** Circumferential strain

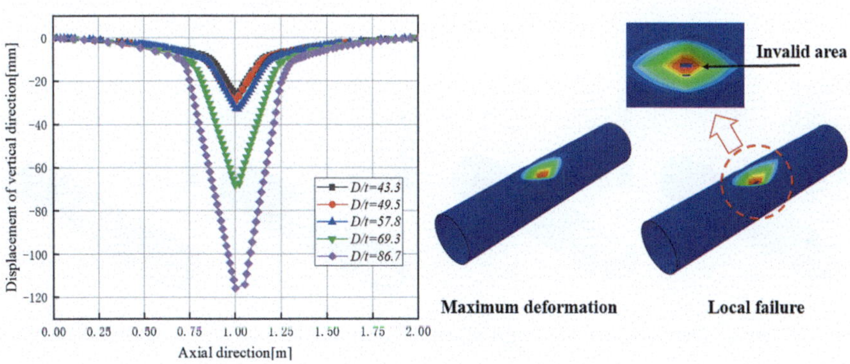

Fig. 7.22 Pipe deformations with different diameter-thickness ratios under spherical indenter load

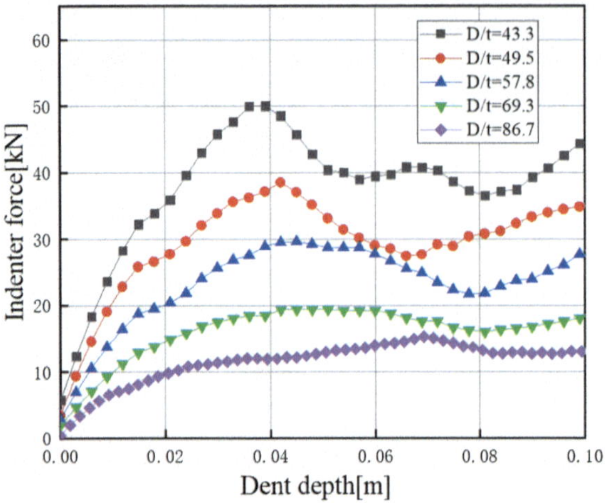

Fig. 7.23 Force–displacement curve with different diameter-thickness ratios under spherical indenter load

Influence of Layup Scheme

Stress and Displacement: Different layup schemes result in varied stress distributions and displacement responses under spherical indenter load. Axial fiber winding ($[0°]_{24}$) leads to uniform loading across layers, while circumferential winding ($[45°/0°/-45°/90°]_6$) results in more localized depression and stress concentration (Figs. 7.24 and 7.25).

Figure 7.26 shows the ultimate load and dent depth of pipe with different layup schemes. Circumferential winding schemes exhibit greater compressive displacement, but lower ultimate loads compared to axial winding, suggesting a trade-off between load-bearing capacity and deformation.

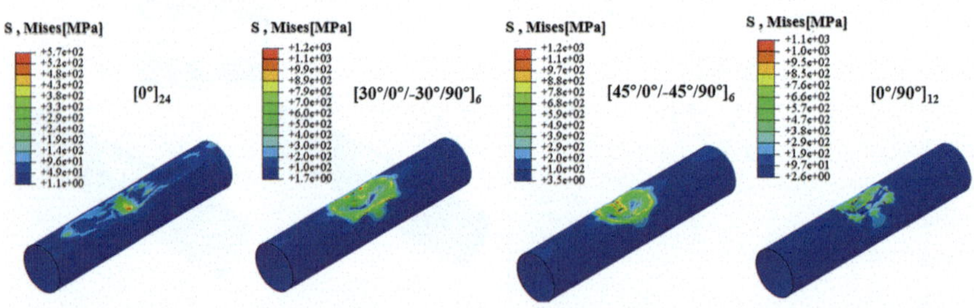

Fig. 7.24 Stress distribution of pipe under different layup scheme under spherical indenter load

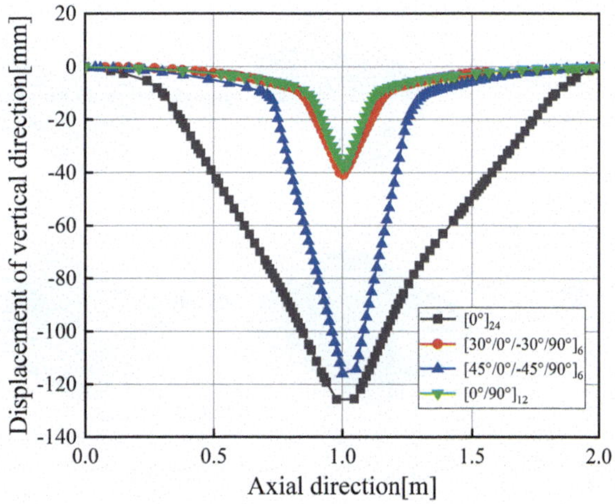

Fig. 7.25 Deformations of the pipe with different layup scheme under spherical indenter load

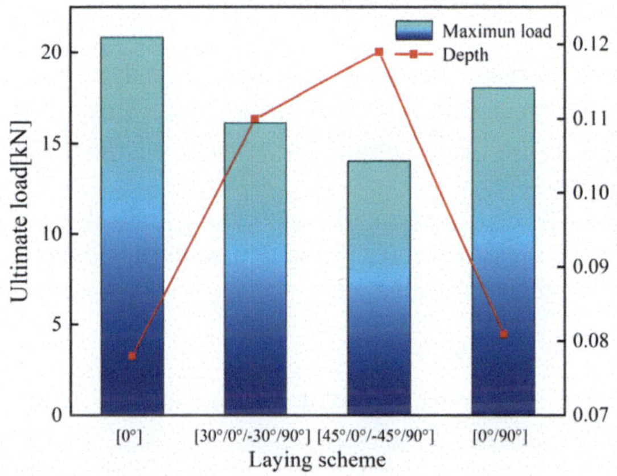

Fig. 7.26 Ultimate load and dent depth of pipe with different layup schemes

7.4.2.2 Plate Load Condition

Stress Distribution and Deformation: Under plate load, BFRP experiences two main deformation stages, visible in Fig. 7.27. Initially, four plastic hinge lines form, enhancing strength and stability. High stress zones appear around these lines. In the second stage, the pipe wall dents and flattens, reducing stress concentration. The cross-section transforms into an arch with a symmetrical shape.

Load and Energy Curves: Fig. 7.28a displays the BFRP's load response, showing initial elasticity, followed by a decrease in stiffness and bearing capacity due to fiber and matrix damage. The remaining undamaged material continues to resist deformation. Figure 7.28b presents the energy absorption curve, correlating energy absorption with bearing capacity.

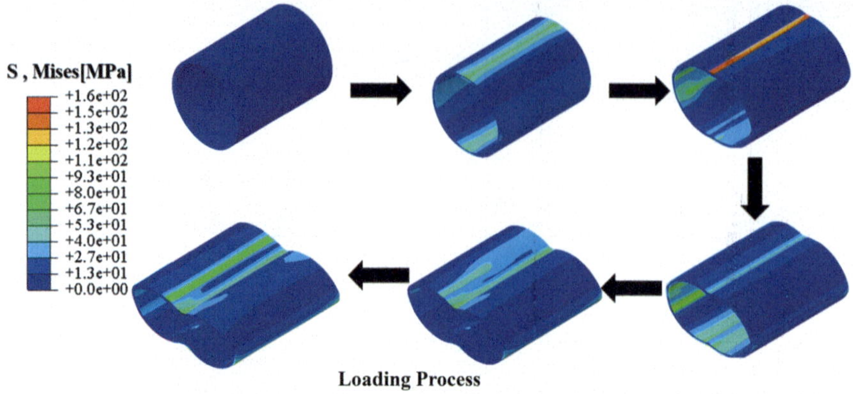

Fig. 7.27 Stress distribution of the pipe under plate load

7.4.2.3 Influence of Internal Pressure

Matrix Damage: Fig. 7.29 illustrates matrix damage under varying internal pressures. Tensile damage occurs mainly in the contact area, while compression damage concentrates at the plastic hinges. Increasing internal pressure complicates the damage pattern.

Load–Displacement and Energy: Fig. 7.30 shows that internal pressure enhances the pipe's resistance to deformation, requiring higher loads for failure and improving energy absorption efficiency.

7.4.2.4 Influence of Layup Scheme and Winding Angle

Fiber Layer Damage: Fig. 7.31 reveals varied damage patterns across different fiber layers under plate load. Axial winding (Ply-0°) experiences less fiber damage, mainly in tensile form. Ply-45° shows complex damage, with compression damage dominating. Ply-90° exhibits damage parallel to the pipe's central axis.

Stress Distribution: Fig. 7.32 demonstrates that deformation under plate load is symmetrical and depends on the winding angle. Increasing the winding angle enhances circumferential stiffness, stabilizing the deformation.

Load–Displacement and Energy with Diameter-Thickness Ratios: Fig. 7.33 indicates that smaller diameter-thickness ratios result in higher load values and improved energy absorption efficiency.

Figure 7.34 shows that larger winding angles increase the pipe's strength, stiffness, and energy absorption efficiency, highlighting the anisotropic nature of basalt fiber composite material.

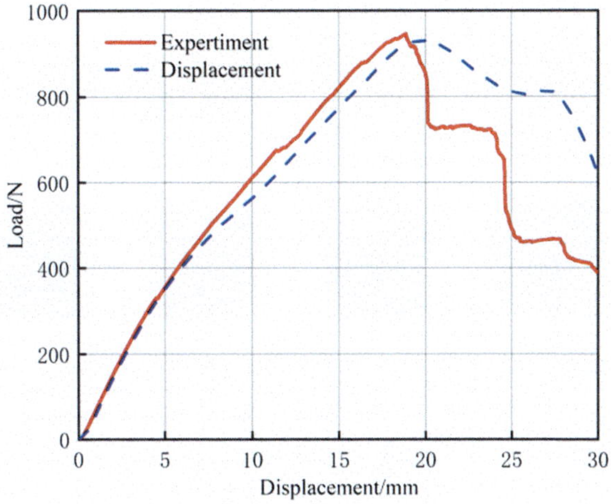

(a) Force-displacement

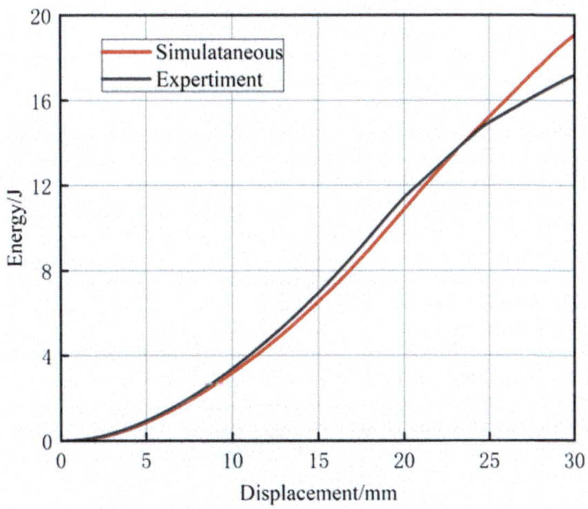

(b)Energy-displacement

Fig. 7.28 Load displacement curve and energy curve of composite pipe under plate load. **a** Force–displacement, **b** energy-displacement

7.5 Conclusions

This chapter has systematically analyzed the performance of fiber-wound composite pipes under diverse load conditions, highlighting the significant roles of winding angle, internal pressure, and diameter-thickness ratio.

Load Type and Winding Angle: The inner layer of the reinforced layer is prone to failure under internal pressure, with larger winding angles enhancing pressure-bearing capacity up to a certain limit. Torsional load failures occur predominantly in the outer layer of the reinforced layer, optimal at winding angles between ± 35°

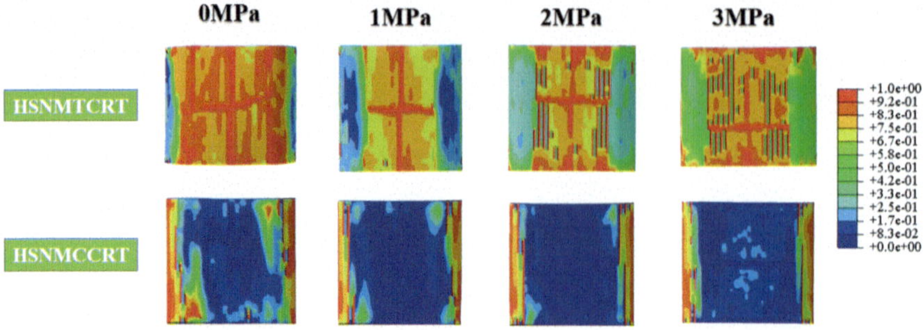

Fig. 7.29 Matrix damage of composite pipe with different internal pressure under plate load

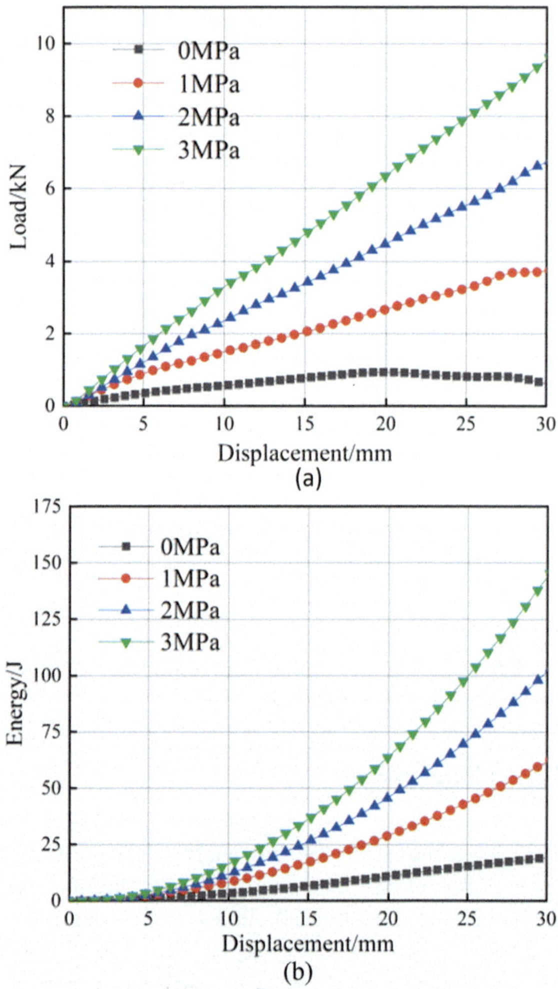

Fig. 7.30 Load–displacement and energy curves with different internal pressures underplate load. **a** Force–displacement, **b** energy-displacement

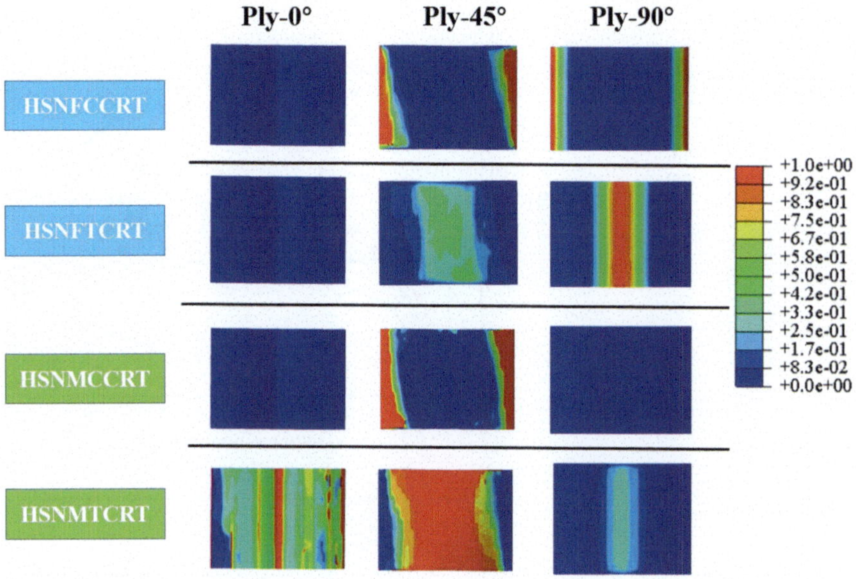

Fig. 7.31 Damage diagram of different fiber layers under plate load

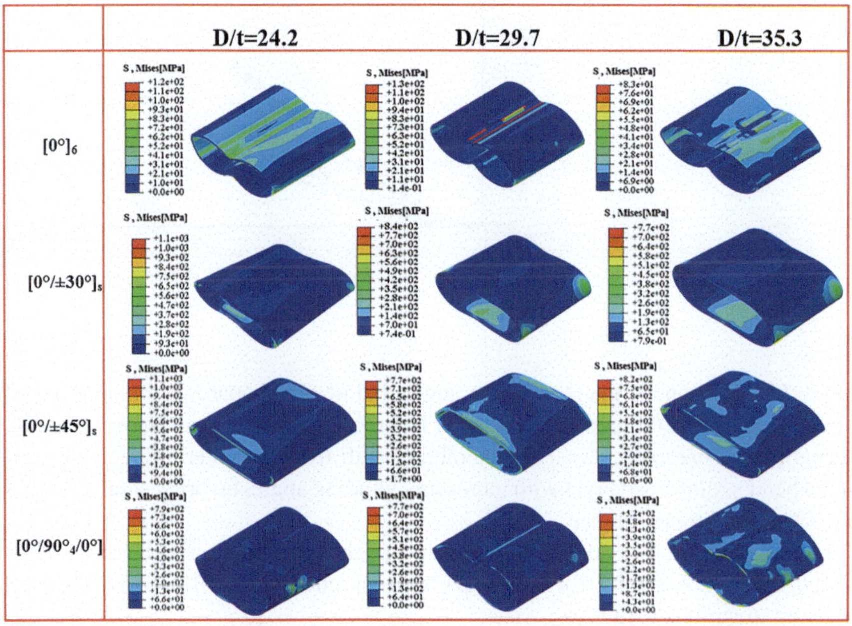

Fig. 7.32 Stress distribution of the pipe with different diameter thickness ratio and winding angle

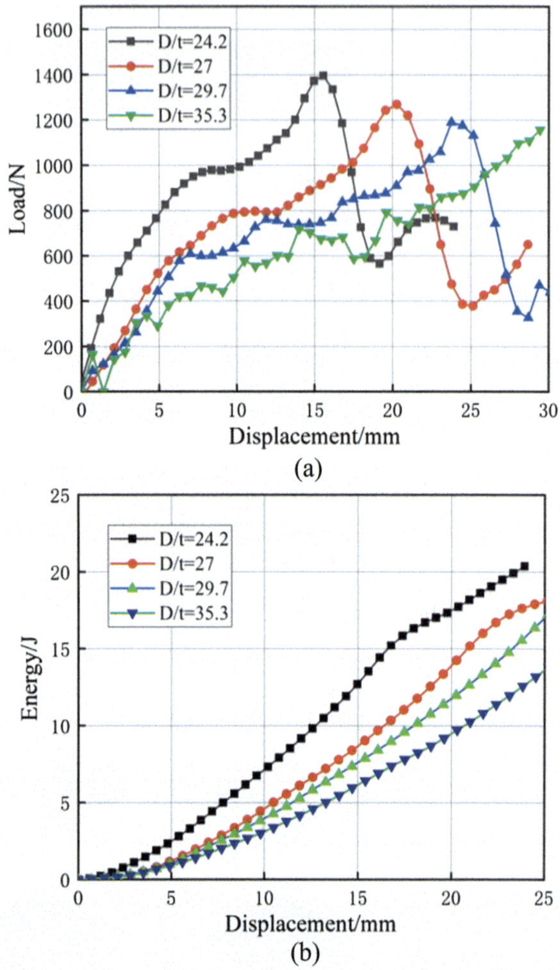

Fig. 7.33 Load displacement and energy curves with different diameter-thickness ratios under plate load. **a** Force–displacement, **b** energy-displacement

and $\pm$ 55°. For tensile loads, winding angles below 60° jeopardize the inner layer, while angles above 60° endanger the outer layer, with smaller angles boosting tensile strength. Bending loads reveal a vulnerability shift from the inner to the outer layer of the bend's reinforced layer with increasing winding angle, favoring smaller angles for better bending resistance.

Optimization and Load-Bearing: A multi-angle winding approach for the reinforced layer is essential for maximizing comprehensive load-bearing capacity, adapting to various load types.

Spherical Indenter and Plate Load Behavior: The pipe's response to a spherical indenter involves an initial force spike, followed by a decrease due to fiber matrix failure, leading to eventual pipe failure. The ultimate load capacity is inversely related to the diameter-thickness ratio, and is influenced by the winding angle. Under

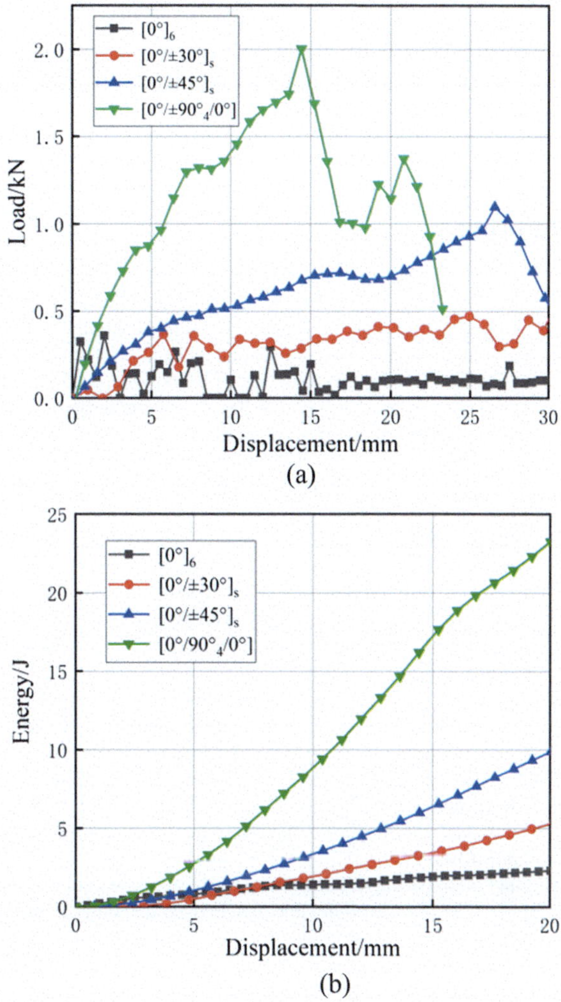

Fig. 7.34 Load displacement and energy curve with different layup scheme under plate load. **a** Force–displacement, **b** energy-displacement

plate load, internal pressure boosts energy absorption efficiency, while the fiber winding angle impacts the failure area and pipe stiffness. Increasing the diameter-thickness ratio reduces both load-bearing capacity and energy absorption efficiency. The complexity of fiber and matrix failure, alongside delamination under external loads, necessitates further research.

In essence, this chapter provides crucial insights for optimizing fiber-wound composite pipe design, ensuring enhanced performance across various load scenarios.

This chapter is adapted from the article ‘Failure factor of reinforced layer for fiber-wound composite pipe under different loads,’ originally published in *Strength, Fracture and Complexity* (2023) DOI: https://doi.org/10.3233/SFC-230012. Reproduced with permission from IOS Press.

References

1. Bazli M, Benny B, Rajabipour A, Pourasiabi H, Heitzmann MT, Arashpour M (2023) Residual compressive strength of seawater sea sand concrete filled hybrid carbon-glass fibre reinforced polymer tubes under seawater: effects of fibre type and orientation. J Build Eng 70:106383. ISSN 2352-7102. https://doi.org/10.1016/j.jobe.2023.106383
2. Liu TQ, Liu X, Feng P (2020) A comprehensive review on mechanical properties of pultruded FRP composites subjected to long-term environmental effects. Compos Part B: Eng 191:107958. ISSN 1359-8368, https://doi.org/10.1016/j.compositesb.2020.107958
3. Tsai SW, Wu EM (1971) A general theory of strength for anisotropic materials. J Compos Mater 5(1):58–80
4. Sargado JM, Keilegavlen E, Berre I, Nordbotten JM (2017) Analysis of Hashin failure criteria for tensile and compression fracture of fibre-reinforced composite using finite element technique. http://arxiv.org/abs/1705.04046. Accessed 16 Oct 2017
5. Xin Z (2020) Study on mechanical properties and optimum design of thermoplastic composite flexible pipe under internal and external pressure. China University of Petroleum (East China), Qingdao
6. Xunyun Z, Jun H, Ziyi X et al (2013) Study on residual strength and remaining life of externally-corroded pipelines based on finite element analysis. Chem Eng Mach 40(05):639–641
7. Rafiee R, Amini A (2015) Modeling and experimental evaluation of functional failure pressures in glass fiber reinforced polyester pipes. Comput Mater Sci 96:579–588
8. Lu Y, Shuqing W, Wentao H et al (2020) Theoretical and numerical analysis of glass-fiber reinforced thermoplastic flexible pipe for marine engineering. Period Ocean Univ China 50(06):141–148
9. Jun P, Dongsheng Z, Peng L et al (2018) Cross section structure design and strength analysis of glass fiber reinforced flexible pipe in deep sea. Fiber Rein-Forced Plast/Compos 2018(02):21–27
10. Shi C, Xia H, Wang J et al (2022) Partially-plastic theoretical model of thermoplastic composite pipes and comparison of composite failure criteria. Compos Struct 280:114834.A

Chapter 8
Dent Behaviors of Steel Pipes Under External Load

8.1 Introduction

This chapter explores the phenomenon of pipeline denting, a significant form of mechanical damage in steel pipes, resulting from external forces that cause permanent deformation and distortion [1]. Dents typically occur during installation or operation due to interactions with rocks or other hard objects, leading to stress concentrations that can precipitate failure [2]. The behaviors of dented pipelines are examined to ensure their integrity and safe operation. The failure mechanisms associated with different dent orientations, the impact of pipe geometry on mechanical responses, and the rebound characteristics of dents are analyzed. The chapter also investigates the various types of dents and their implications for pipeline performance. Through a combination of empirical and numerical analyses, insights into the complexities of dent behavior are provided, informing maintenance, repair, and failure prevention strategies for steel pipelines.

8.2 Numerical Simulation Model

In engineering, pipes may encounter various external contacts, such as longitudinal, transverse, or tilted, leading to different dent shapes [3, 4]. Figure 8.1 illustrates a triangular prism indenter with a 100-mm radius arc contacting a pipe with a 508 mm diameter and a length six times that. The simulation used refined meshing near the contact area, employing four-node shell elements for the pipe and eight-node elements for the indenter. The study utilized a large-strain plasticity model with isotropic hardening for the steel pipe, specifically using X65 steel—a common material in oil and gas pipelines. The material's properties include a yield stress of 448 MPa, Young's modulus of 210GPa, Poisson's ratio of 0.3, and density of 7800 kg/m^3. Simulations applied a displacement load to the indenter, which was increased

H. Zhang et al., *Submarine Pipeline Integrity: Assessment of Failure Modes and Advanced Evaluation Techniques*, Ocean Engineering & Oceanography 28,
https://doi.org/10.1007/978-3-031-92092-9_8

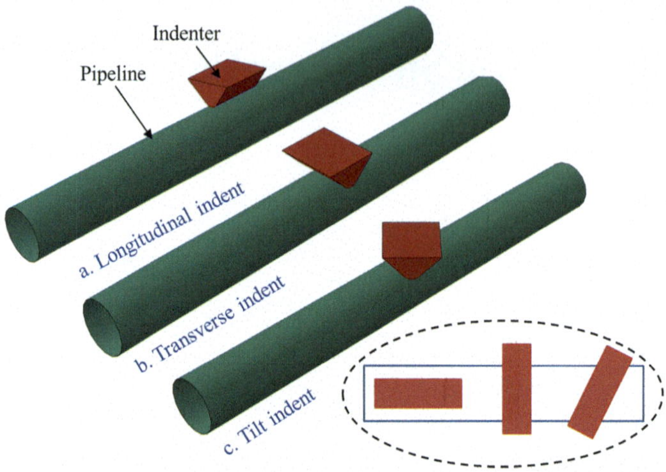

Fig. 8.1 Schematic diagram of the calculation model

to a predefined maximum and then reduced to zero, encompassing both the loading and unloading phases.

Reference [5] details experiments using a hardened steel indenter with a 90° head angle and a 2 mm radius to apply a local transverse load on pipe specimens. These pipes measured 2 m in length, 140 mm in diameter, and had a wall thickness of 2.555 mm. Deformation results, depicted in Fig. 8.2, reveal a "V" shaped dent and squashing at the pipe's lower section due to the rigid base. Figure 8.3 compares load–displacement curves from experimental and simulation tests, showing minimal discrepancy, thus validating the numerical model's reliability despite some errors in the load–displacement data.

8.3 Results and Discussion

8.3.1 Mechanical Behaviors

Under longitudinal denting, the pipe experienced deformation with an initial indenter displacement of 0.16 m, leading to a dent that deepened as the indenter pressed further, as illustrated in Fig. 8.4. Upon unloading, some elastic rebound was observed, but permanent plastic deformation remained, resulting in a flat-bottomed dent shaped by the columnar indenter's profile.

Figure 8.5 depicts the force–displacement relationship during longitudinal denting of a pipe, characterized by five distinct stages: initial elastic behavior (before K_1), onset of plastic deformation ($K_1 \sim K_2$), a plateau where dent depth increases without

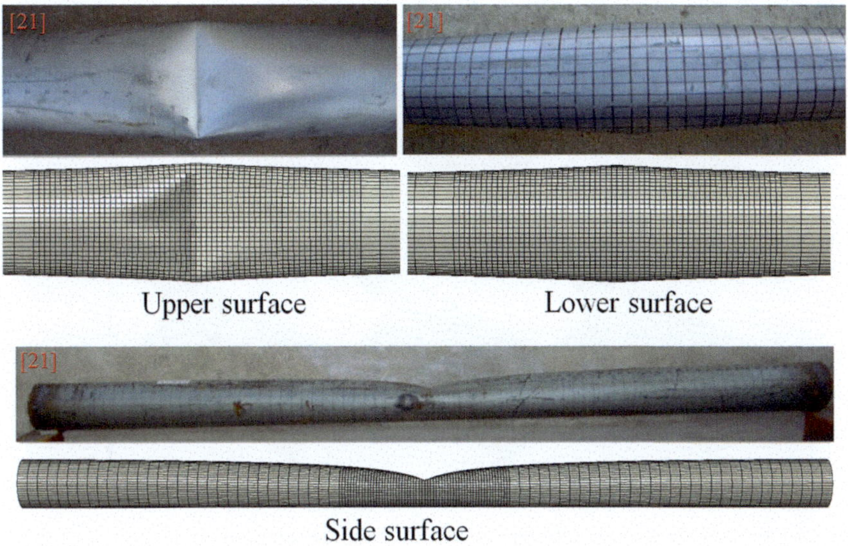

Fig. 8.2 Comparison of pipe deformation obtained from experimental tests and numerical simulations

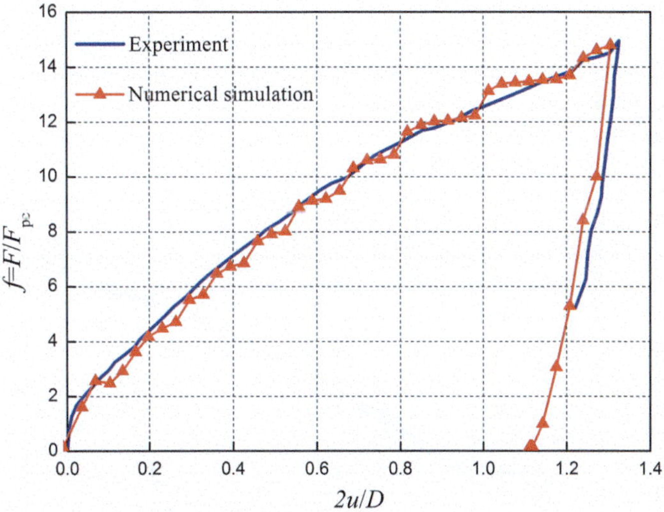

Fig. 8.3 Normalized load–displacement curves of the pipe obtained from experimental and numerical simulations

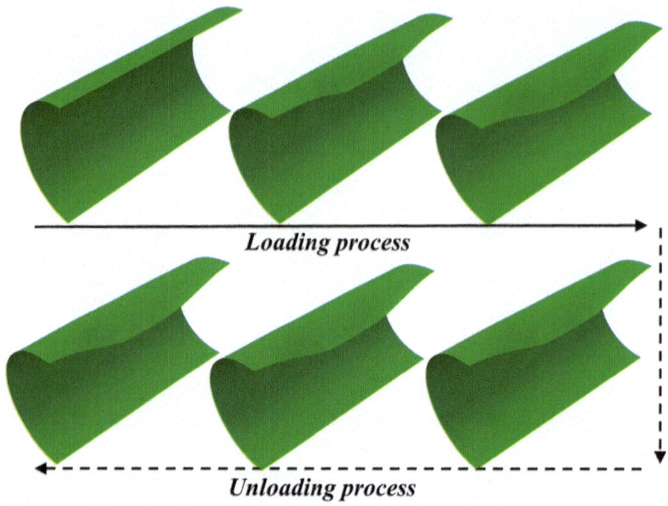

Fig. 8.4 Deformation process of the pipe undergoing longitudinal denting

additional force (K_2 ~ K_3), a nonlinear increase in force with further denting (K_3 ~ K_4), and finally, an unloading phase with elastic rebound leaving a permanent dent.

For transverse denting, shown in Fig. 8.6, the pipe exhibits a V-shaped dent and a force–displacement curve (Fig. 8.7) that differs from longitudinal denting, with a less pronounced plateau and a combination of elastic rebound and plastic deformation during unloading.

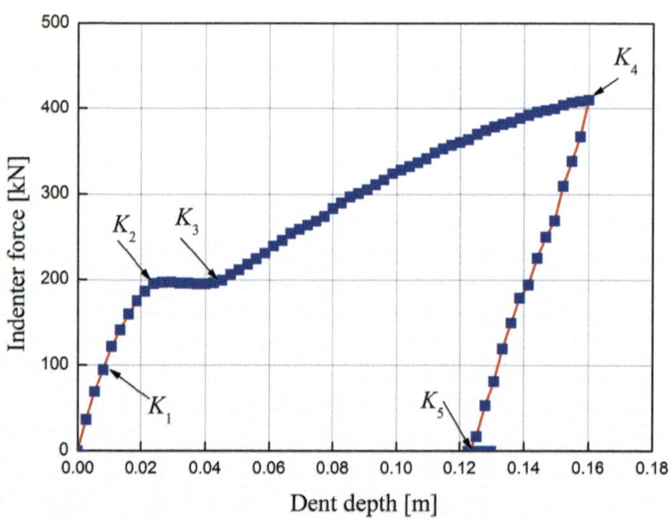

Fig. 8.5 Force–displacement curve undergoing longitudinal denting

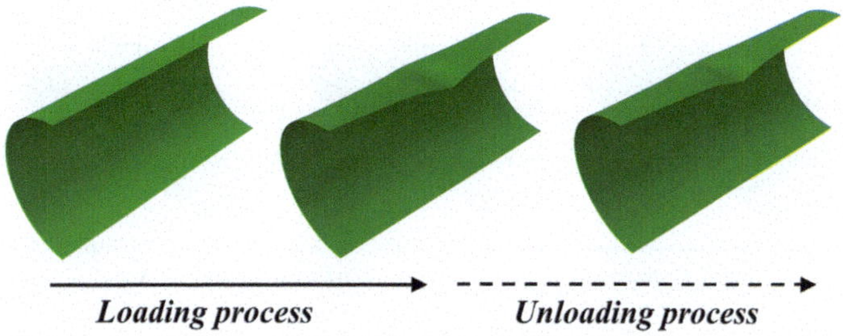

Fig. 8.6 Deformation process of the pipe undergoing transverse denting

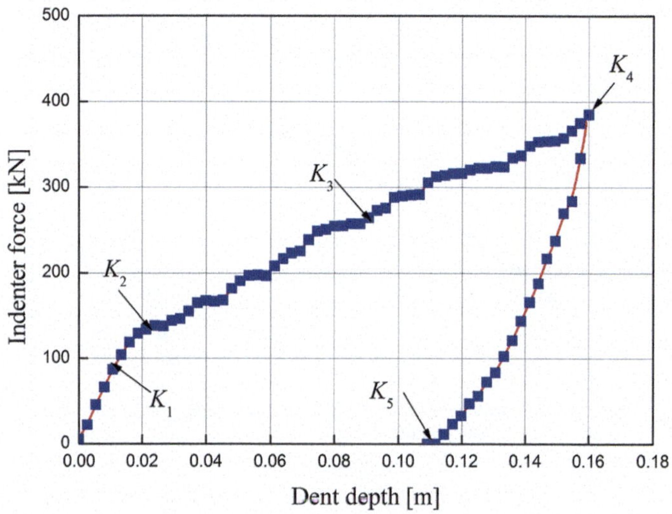

Fig. 8.7 Force–displacement curve of the pipe undergoing transverse denting

Tilted denting, presented in Fig. 8.8, results in an oval cross-section near the dent. Its force–displacement curve (Fig. 8.9) lacks a plateau phase and includes four stages: elastic deformation, initial plastic deformation, progressive plastic deformation, and a rebound stage, with the unloading curve indicating both elastic and plastic behavior.

8.3.2 Effect of Different Indenter Orientation

8.3.2.1 Longitudinal Denting

Figure 8.10 illustrates the von Mises stress distribution in a pipe during different denting stages. Initially, high stress is localized at the contact point between the indenter and the pipe, spreading axially and circumferentially with increased indenter

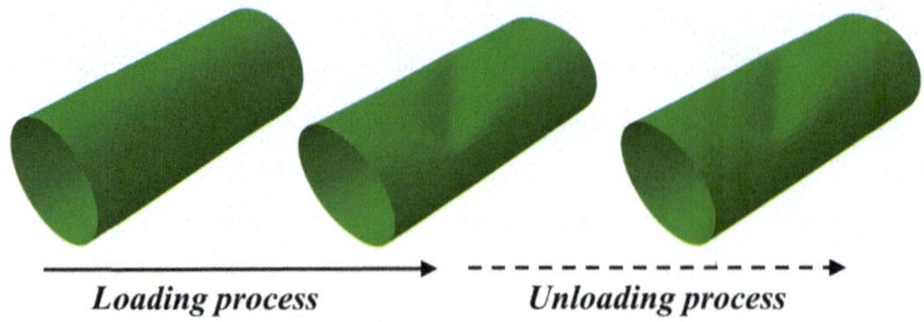

Fig. 8.8 Deformation process of the pipe undergoing tilted denting

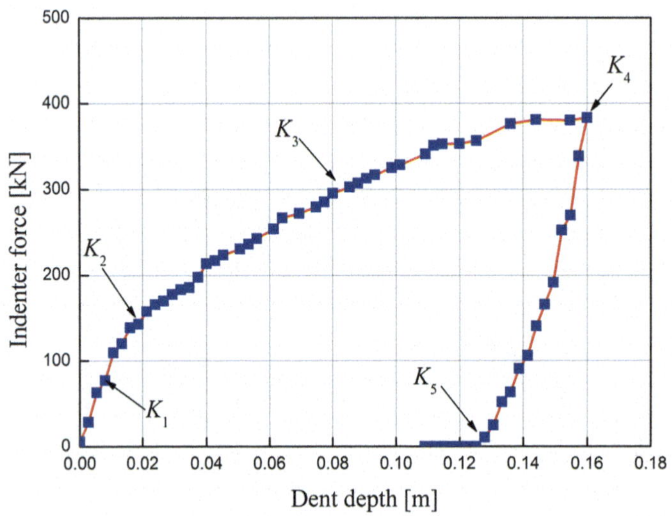

Fig. 8.9 Force–displacement of the pipe undergoing tilted denting

displacement. Maximum stress is observed in the dent area at peak indentation, which diminishes upon unloading, concentrating on the dent's outer edge.

Figure 8.11 reveals the equivalent plastic strain during longitudinal denting, concentrated at the dent with the highest strain at the indenter's ends. Initially, equivalent strains at these points are equal, indicating no significant dent. As the dent deepens, strain increases, with the middle point's rate of increase slowing and the end point's strain stabilizing after a certain depth. Unloading also increases strain, but it stabilizes after elastic rebound.

Figure 8.12 shows the axial strain distribution, with compressive strain at the dent and tensile strain on either side. The maximum compressive strain, caused by the indenter's sharp ends, suggests potential crack formation at the dent bottom, the most vulnerable area. The axial strain profile along the pipe's length exhibits a 'raised effect' due to the cylindrical shell's resistance.

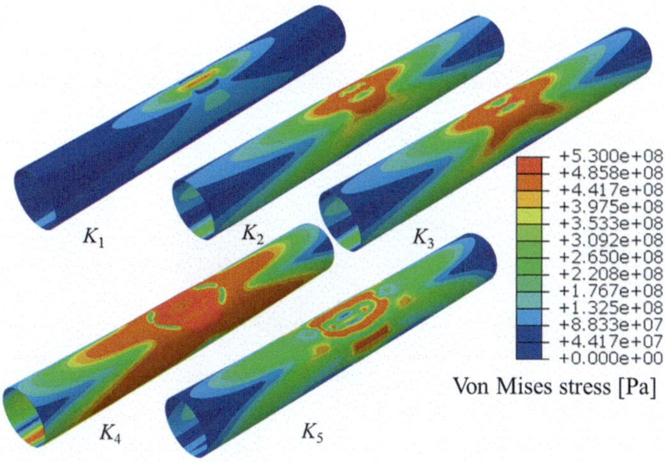

Fig. 8.10 Stress distribution of the pipe undergoing longitudinal denting

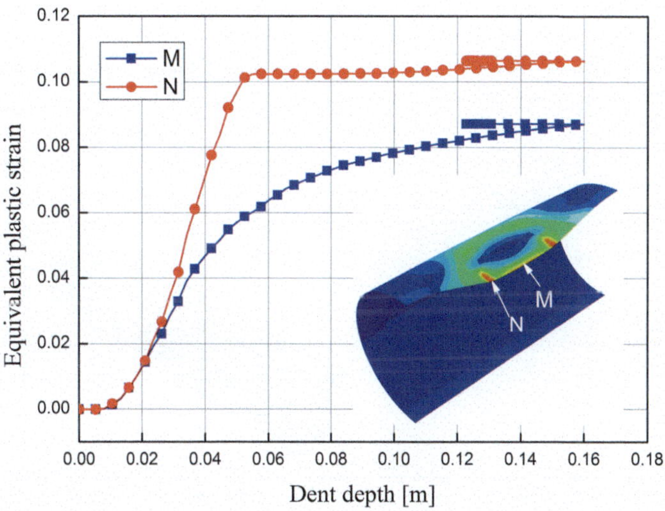

Fig. 8.11 Equivalent plastic strain of the pipe undergoing longitudinal denting

8.3.2.2 Transverse Denting

Figure 8.13 depicts the pipe's longitudinal deformation due to transverse denting, mirroring the columnar indenter's cross-section. At the symmetrical axis, deformation remains consistent from the limit to the final state, with a rebound rate of about 29.65%.

Figure 8.14 details the pipe's axial strain distribution under transverse denting, showing higher strain at the dent with compressive strain at the bottom and higher

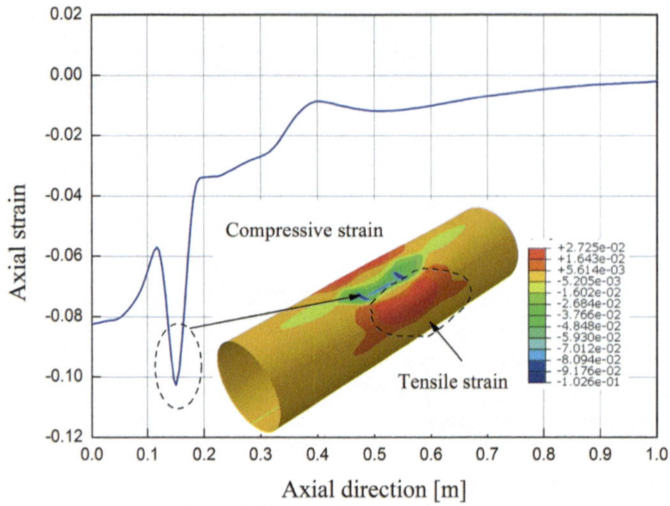

Fig. 8.12 Axial strain distribution of the pipe undergoing longitudinal denting

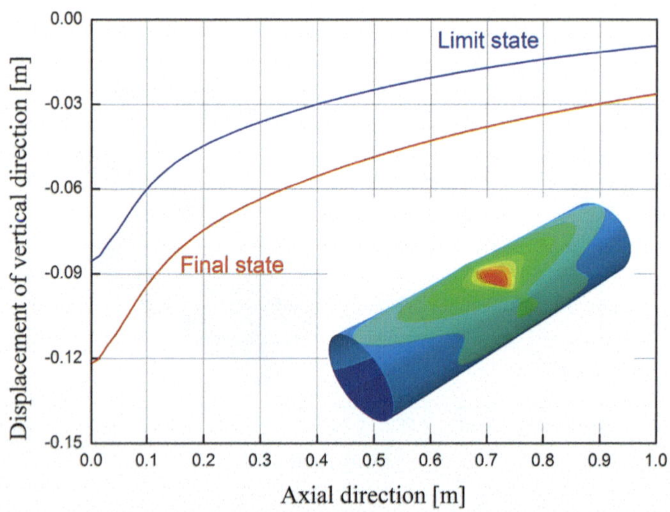

Fig. 8.13 Deformation of the pipe undergoing transverse denting

tensile strain at the sides, indicating a higher failure propensity at the sides compared to the dent center, contrasting with longitudinal denting behavior.

Figure 8.15 illustrates the plastic deformation, with the highest plastic strain at the dent's sides, not the center. Plastic deformation initiates at the top and increases with dent depth, stabilizing after 0.05 m. At the M′-point, strain escalates rapidly below

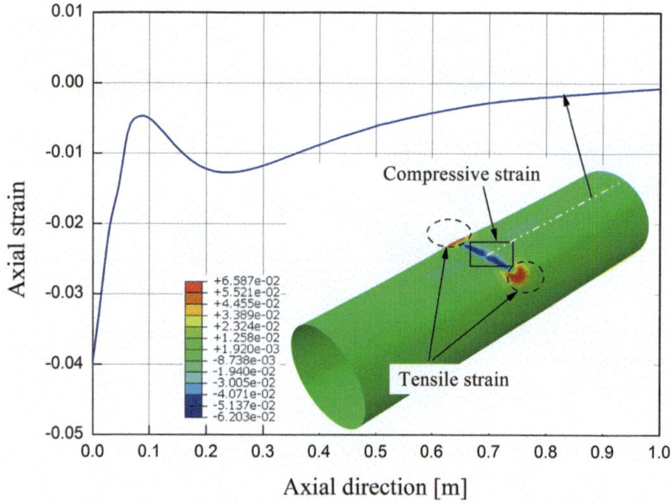

Fig. 8.14 Axial strain of the pipe undergoing transverse denting

0.11 m dent depth, surpassing the N′-point strain beyond 0.078 m. During unloading, the N′-point strain remains stable, while the M′-point strain varies significantly.

Figure 8.16 shows the von Mises stress evolution under transverse denting, starting high at the top and spreading circumferentially and axially during loading, transitioning from oval to star, then hexagonal shapes. Post-unloading, stress diminishes except at the dent, indicating a stress concentration and residual weakness at the dent location.

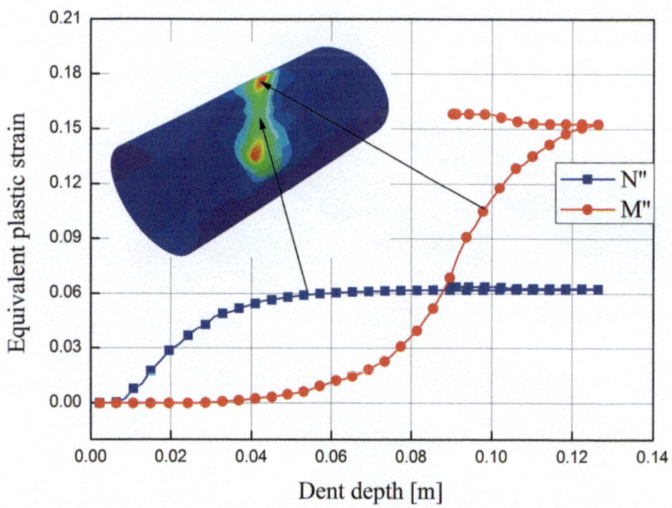

Fig. 8.15 Equivalent plastic strain of the pipe undergoing transverse denting

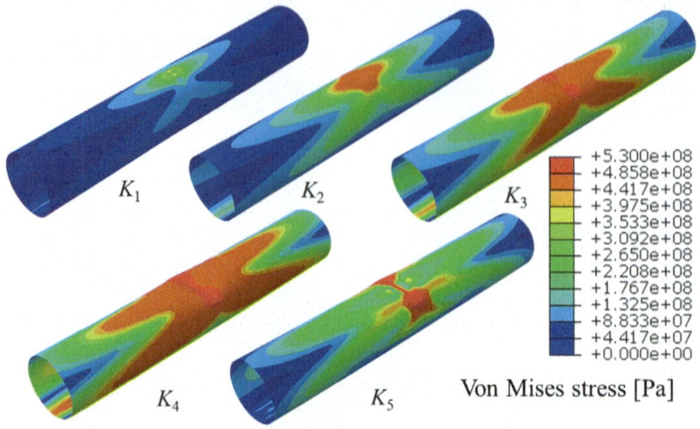

Fig. 8.16 Von Mises stress of the pipe undergoing transverse denting

8.3.2.3 Tilted Denting

Figure 8.17 illustrates the von Mises stress distribution in a pipe experiencing tilted denting, showing antisymmetric stress around the center. Initially, high stress is localized at the indenter contact, expanding axially and circumferentially with indenter movement. The stress peaks at the dent's bottom and, despite a post-unloading decrease, remains concentrated there and more severe at the sides.

Figure 8.18 presents the pipe's plastic deformation and strain under tilted denting, with deformation starting in the middle and spreading to the sides. Plastic strain increases even after unloading.

Figure 8.19 details the axial strain distribution, indicating the dent location as the most vulnerable due to higher compressive than tensile strain. The strain

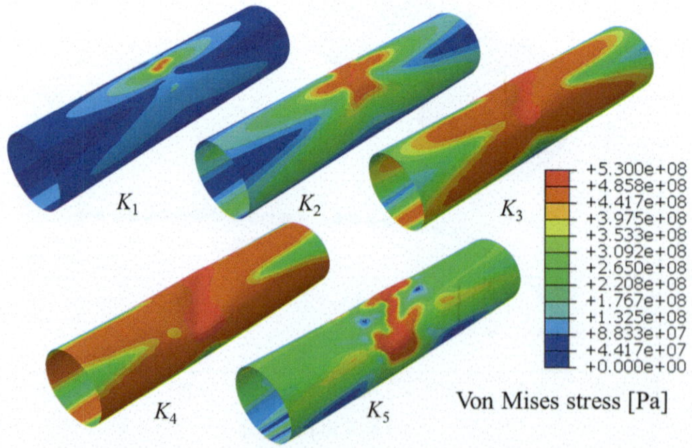

Fig. 8.17 Von Mises stress of the pipe undergoing tilted denting

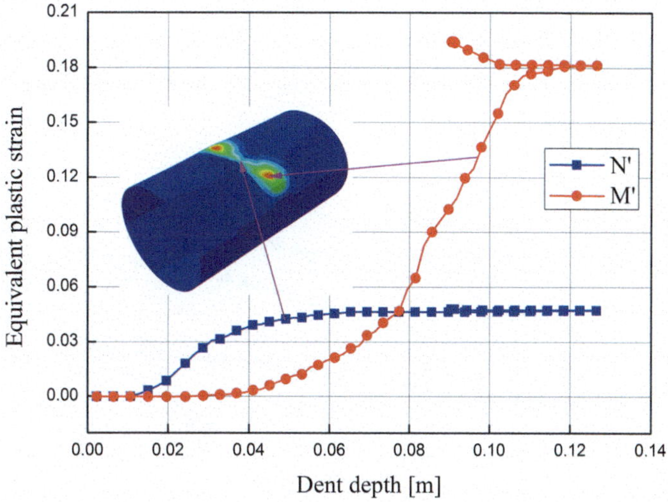

Fig. 8.18 Equivalent plastic strain of the pipe undergoing tilt denting

curve confirms the maximum strain at the dent bottom, differentiating tilted denting behavior from longitudinal and transverse denting.

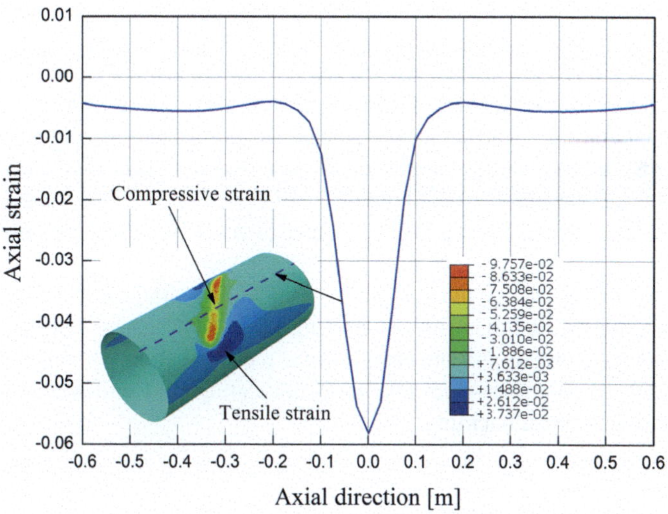

Fig. 8.19 Axial strain distribution of the pipe undergoing tilt denting

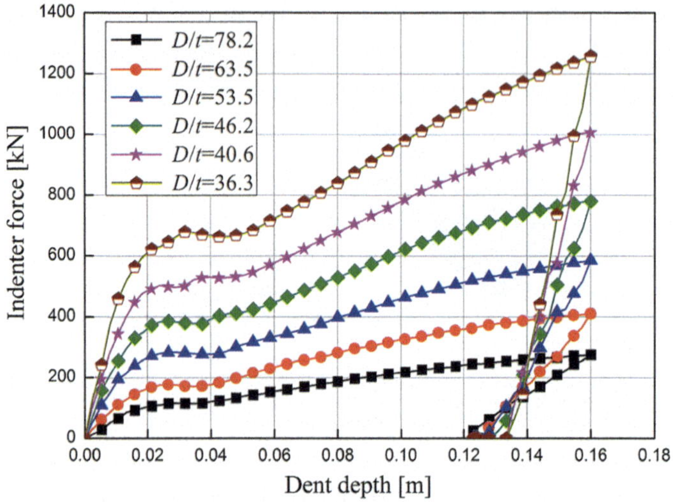

Fig. 8.20 Force–displacement curves of the pipe with different diameter-thickness ratios

8.3.3 *Effect of Essential Parameters*

8.3.3.1 Diameter-Thickness Ratio

The stiffness of a pipe is influenced by its diameter-thickness ratio; a smaller ratio results in higher stiffness. As shown in Fig. 8.20, with a constant dent depth, the force required to dent the pipe rises as the diameter-thickness ratio decreases, and the force–displacement curve's plateau is more pronounced for pipes with smaller ratios. Upon unloading, steeper curve slopes are observed with smaller ratios, indicating a higher likelihood of denting in thin-walled pipes.

Figure 8.21 reveals that for a given indenter displacement, thinner pipes exhibit deeper dents, suggesting a higher susceptibility to external forces.

According to Fig. 8.22, thicker pipes, with higher diameter-thickness ratios, show reduced plastic and axial strains, making them safer against external impacts. The rebound rate, or the proportion of elastic recovery to maximum dent depth, increases with the diameter-thickness ratio, indicating that thin-walled pipes are more vulnerable to local buckling.

8.3.3.2 Initial Indenter Displacement

The rebound rate of a pipe, as depicted in Fig. 8.23, diminishes with greater initial indenter displacement due to more pronounced local buckling that resists elastic recovery. This relationship between indenter force and displacement is nonlinear, with the force rising at a decreasing rate as displacement increases.

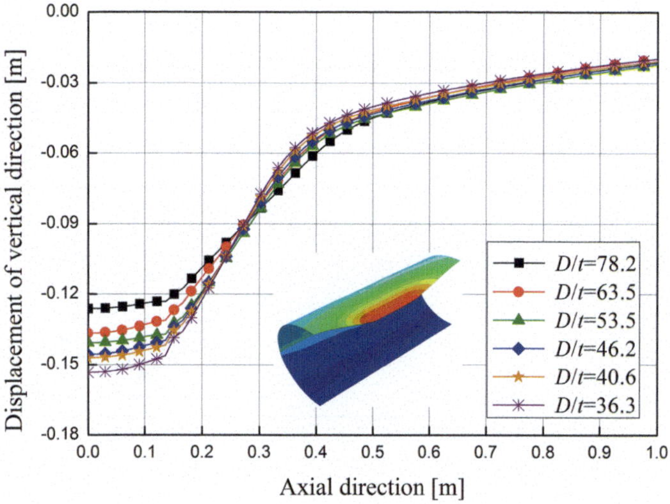

Fig. 8.21 Deformations of the pipe with different diameter-thickness ratios

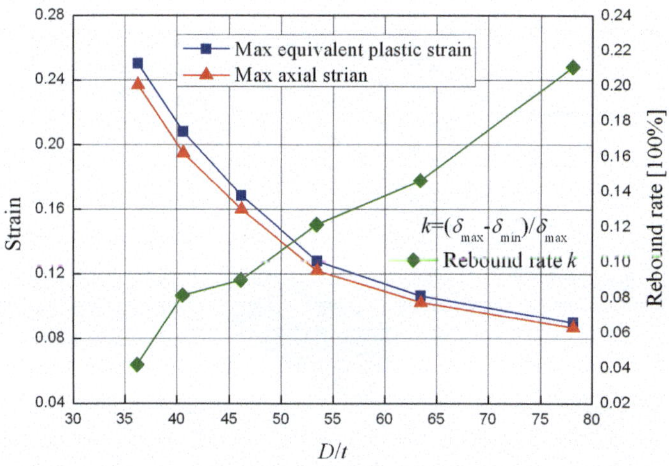

Fig. 8.22 Strain and rebound rate of the pipe with different diameter-thickness ratio

Figure 8.24 illustrates that larger initial indenter displacements lead to more extensive permanent plastic deformation in pipes, with a longer transition from the dent to the undamaged section. Smaller displacements result in flatter dent bottoms.

As shown in Fig. 8.25, the maximum axial strain escalates with indenter displacement, which marginally affects the plastic deformation zone. Below a displacement-to-diameter ratio (w/D) of 0.4, the peak plastic strain occurs at the dent bottom. Above this ratio, the maximum strain shifts to the sides of the dent.

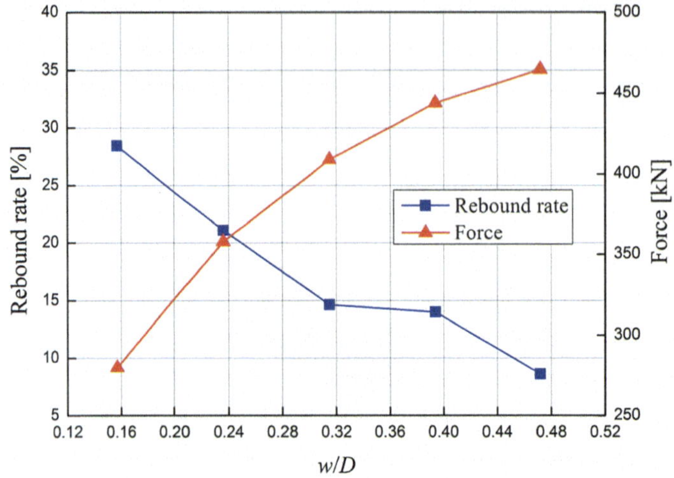

Fig. 8.23 Force and rebound rate of the pipe under different indenter displacements

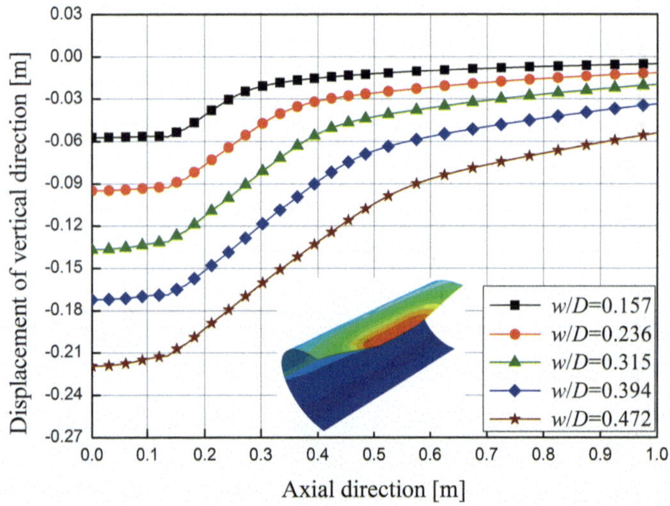

Fig. 8.24 Deformations of the pipe under different indenter displacements

8.4 Conclusions

The chapter has provided a comprehensive analysis of steel pipe failure due to denting, examining the mechanical behavior under longitudinal, transverse, and tilted dents. The investigation into the effects of diameter-thickness ratio and indenter displacement has yielded several key findings:

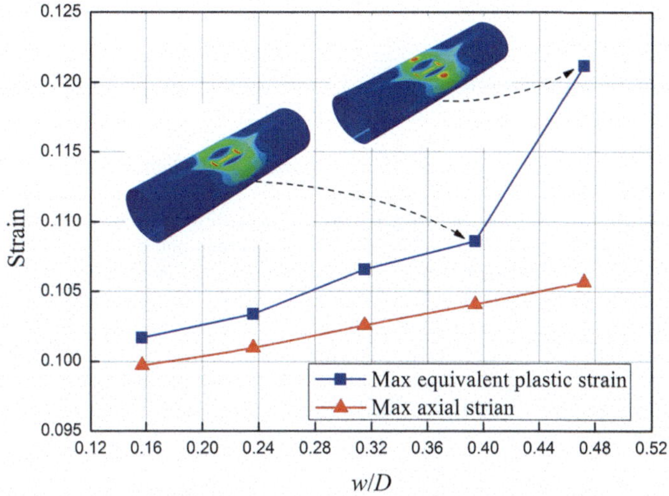

Fig. 8.25 Strain of the pipe under different indenter displacements

Longitudinal denting of pipes reveals a force–displacement curve that can be segmented into five distinct stages, characterized by stress concentration and plastic deformation concentrated at the dent, particularly where the pipe contacts the indenter ends. These areas are most susceptible to failure. Notably, the pipe's rebound rate improves with a higher diameter-thickness ratio, while the axial strain and plastic strain escalate with greater indenter displacement. However, the rebound rate diminishes with increased initial indenter displacement, indicating resistance to elastic recovery due to more severe local buckling.

Transverse denting also presents a five-stage force–displacement curve, with the maximum plastic strain located at the sides of the dent, making these areas particularly vulnerable. The unloading process is defined by a curved force–displacement relationship, suggesting a combination of elastic rebound and permanent plastic deformation.

Tilted denting displays a unique force–displacement curve without a plateau stage, divided into four stages. The stress distribution is antisymmetric around the pipe's center, with the highest residual stress at the dent's bottom. Plastic deformation initiates in the pipe's middle and progresses to the sides of the dent.

These insights into the denting behavior of steel pipes under various conditions are crucial for predicting failure modes and enhancing the design and maintenance strategies to improve the resilience and safety of pipeline systems.

This chapter is adapted from the article ' Simulation investigation of dent behavior of steel pipe under external load,' originally published in *Engineering Failure Analysis* (2018). DOI: https://doi.org/https://doi.org/10.1016/j.engfailanal.2018.03.024. Reproduced with permission from Elsevier.

References

1. Allouti M, Schmitt C, Pluvinage G, Gilgert J, Hariri S (2012) Study of the influence of dent depth on the critical pressure of pipeline. Eng Fail Anal 21:40–51. https://doi.org/10.1016/j.engfailanal.2011.11.011
2. Kyriakides S, Corona E (2007) Mechanics of offshore pipes: volume 1 buckling and collapse. Elsevier
3. Junqiang W, Yi S, Can F, Zhang P, Wang T, Lin N, Liu Z (2021) Multi-dimensional mechanical response of multiple longitudinally aligned dents on pipelines and its effect on pipe integrity. Thin-Walled Struct 166:108020. https://doi.org/10.1016/j.tws.2021.108020
4. Cosham A, Hopkins P (2004) The effect of dents in pipelines—Guidance in the pipeline defect assessment manual. Int J Press Vessels Pip 81(2):127–139. https://doi.org/10.1016/j.ijpvp.2003.11.004
5. Eyvazinejad Firouzsalari S, Showkati H (2013) Behavior of pre-compressed tubes subjected to local loads. Ocean Eng 65(Complete):19–31. https://doi.org/10.1016/j.oceaneng.2013.03.003

Chapter 9
Innovative Techniques in Submarine Pipeline Repair

9.1 Introduction

The maintenance and repair of these pipelines are crucial for ensuring a continuous supply of energy and for protecting marine ecosystems from potential leaks and environmental harm [1]. This highlights the critical nature of submarine pipeline repair, which demands a careful balance of engineering expertise and environmental stewardship. The challenges presented by the underwater environment, such as high pressures, corrosive conditions, and isolated locations, call for the development and implementation of advanced repair techniques that surpass those used for onshore pipelines [2].

This chapter explores the complex world of submarine pipeline repair, beginning with an examination of its importance and the unique challenges of the marine environment. In Sect. 9.2, "Key Repair Techniques," we review various repair methods, ranging from traditional approaches to state-of-the-art technologies, aimed at addressing corrosion and structural issues. Moving to Sect. 9.3, "Emerging Technologies and Future Trends," the focus shifts to groundbreaking solutions including robotics, autonomous underwater vehicles, innovative materials, and digital twin technology. These advancements are poised to transform repair methodologies by increasing efficiency, effectiveness, and adherence to environmental regulations.

The chapter concludes by weaving together these insights to present a forward-looking perspective on how technological innovations can bolster the resilience and reliability of submarine pipeline systems. This contributes to the sustainability and security of global energy supply chains, underscoring the pivotal role of technological progress in the future of submarine pipeline maintenance.

H. Zhang et al., *Submarine Pipeline Integrity: Assessment of Failure Modes and Advanced Evaluation Techniques*, Ocean Engineering & Oceanography 28,
https://doi.org/10.1007/978-3-031-92092-9_9

9.2 Key Repair Techniques

9.2.1 Above-Water Repair

Above-water repair is an effective technique for mending damaged submarine pipelines by lifting the damaged segments above water to perform cutting, removal, and welding of new pipe sections [3]. Suitable for shallow waters up to 20 m deep, it employs specialized engineering vessels for execution. This approach offers rapid repair times, minimal specialized equipment needs, and superior repair quality. Nonetheless, it requires favorable weather conditions and careful management of potential risks, like equipment malfunctions or abrupt weather shifts, to prevent further damage to the pipeline, making it selectively applicable. The detailed process of the lifting repair method, as shown in Fig. 9.1, is presented as follows:

Step 1: Underwater trenching–Perform trenching around the pipeline based on simulation calculations to expose the required length for lifting.
Step 2: Cutting–Use underwater cold cutting to remove the damaged pipeline section after trenching and mud removal.
Step 3: Lifting and Welding–Position flotation bags and cranes as per the lifting calculation report. Alternately lift three points to elevate the pipe ends for flange welding. After inspection, reverse the pipeline end back to the seabed. Repeat for the other end.
Step 4: Underwater Measurement–Conduct precise underwater measurements of the spatial position and orientation between flanges for accurate prefabrication of the replacement sections on the construction vessel's deck.
Step 5: Underwater Docking–Prefabricate the replacement sections based on measurements and complete the underwater docking with divers connecting the replacement segments.

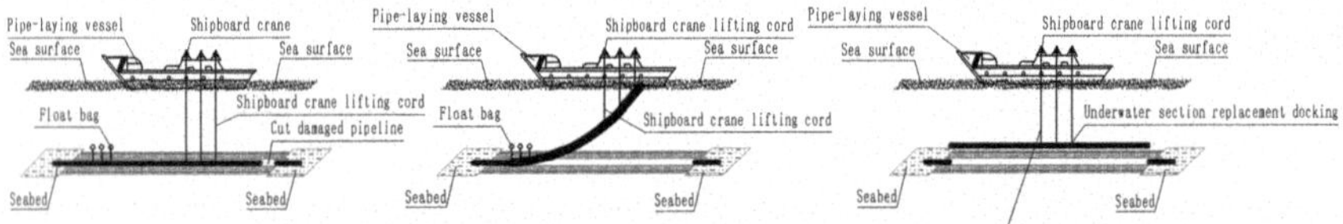

Fig. 9.1 The process of the lifting repair method

9.2.2 *Underwater Maintenance*

9.2.2.1 Clamp Repair

Clamp repair is a targeted solution for repairing minor pipeline damages, such as cracks and corrosion perforations, in low-pressure and low-safety grade underwater pipelines. It involves the strategic placement of a leak-sealing clamp on the affected pipeline section, using external mechanical joints designed for easy underwater installation by divers or ROVs. These clamps feature a dual-structure system equipped with installation rings and a hydraulic mechanism to facilitate seamless opening, closing, and sealing actions [4]. An example of clamp repair for a submarine pipeline is shown in Fig. 9.2. Ensuring pipeline integrity and operational pressure capacity, the clamp's design prioritizes ease of installation and a robust seal, necessitating a smooth pipe surface by removing any external coatings. This method is efficient for localized mechanical damages, offering a practical repair solution for maintaining pipeline functionality.

For more severe damages, encapsulation clamps offer a robust solution by completely encasing the damaged area, providing a comprehensive repair. Split sleeve clamps, known for their ease of installation, are invaluable for fixing cracks or leaks, with their two-piece design allowing for a straightforward wrap-around application without the need for underwater welding. Hot tap clamps facilitate modifications such as new connections or valve insertions on active pipelines, maintaining system pressure throughout the process. Structural reinforcement clamps address areas weakened by corrosion or external impacts, bolstering the pipeline's durability. Composite wrap clamps, leveraging advanced composite materials, serve dual purposes of reinforcing structural integrity and sealing leaks, offering a flexible repair option [5]. Lastly, emergency leak repair clamps are designed for rapid deployment to quickly contain leaks and minimize environmental and operational impacts. Each clamp type is tailored to specific repair needs, factoring in the damage extent, operational conditions, and environmental sensitivities of the repair site, illustrating the nuanced approach required in submarine pipeline maintenance.

Fig. 9.2 Clamp repair for a submarine pipeline

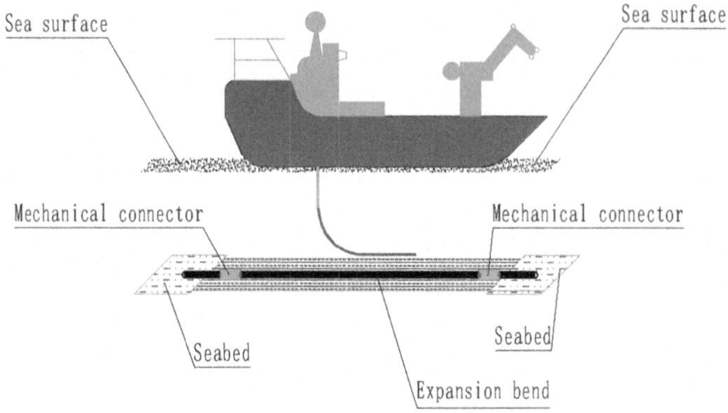

Fig. 9.3 Schematic diagram of mechanical connector installation

9.2.2.2 Mechanical Connector Repair

The mechanical connector repair method revolutionizes the maintenance and repair of submarine pipelines by offering a swift, efficient, and reliable solution for mending damaged sections. This method employs mechanical connectors—innovative devices engineered to seamlessly join two pipeline segments without necessitating welding [6]. These connectors, adaptable for use by divers or remotely operated vehicles (ROVs), prove indispensable in deep-sea or challenging conditions where traditional welding is infeasible. Available in various configurations like bolted flanges, clamp connectors, and couplings, they are designed to accommodate different damage types and pipeline specifications. The schematic diagram of mechanical connector installation is shown in Fig. 9.3. This method's standout advantages include its rapid deployment capability and flexibility, ensuring minimal operational disruption and eliminating the risks associated with welding, such as heat-affected zones. Consequently, it upholds the structural integrity of repairs. Particularly effective for sealing leaks, addressing cracks, or rectifying other damage types, the Mechanical Connector Repair Method is integral to maintaining the operational reliability and safety of critical underwater pipeline infrastructures essential for transporting vital resources across extensive maritime distances. Its capacity for large-area damage repair, applicability across diverse marine environments and depths without specialized ships or welding verification, and the provision of durable, reliable repair solutions underscore its significance in the field.

9.2.2.3 Flanged Repair Method

The flanged repair method is one of the most effective techniques for repairing submarine pipelines with massive damage by replacing the long-distance pipeline with remarkable precision. In case of damage, this method involves mechanical

Fig. 9.4 Ball flange connector [8]

cutting of the respective pipeline and mechanical installation of flanged connectors on these cut ends. Replacement piped sections, pre-welded with flanges at their ends, are then mechanically connected to the original pipeline via these flanged connectors [7]. This method ensures a leak-proof repair while allowing adjustments in angle and installation direction underwater, facilitating alignment corrections between pipeline ends. Figure 9.4 shows a ball flange connector with a flexible connection angle. These include standard flanges, rotary flanges, spherical flanges, and smart flanges. For instance, a spherical flange can compensate for up to 10° of axial misalignment within the flange plane. The design of flanged connectors ensures superior sealing performance by allowing controlled plastic deformation of the pipe wall, thereby achieving a perfect seal and ensuring continuous and efficient underwater pipeline operation. This approach's adaptability to different pipe sizes and resistance to bending moments under operational pressures make it a preferred choice for critical underwater pipeline repairs. However, minor leaks may occur due to imperfections in the mechanical seal, posing a challenge that requires precise installation.

9.2.2.4 Underwater Welding Repair Method

The underwater welding repair method for submarine pipelines is a common approach that diverges significantly from mechanical repair techniques by utilizing

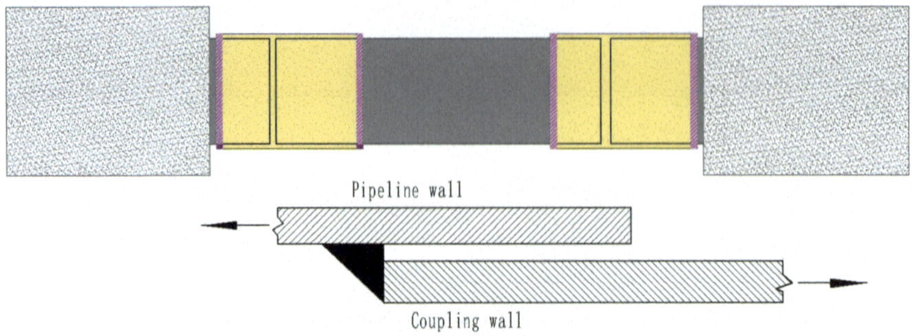

Fig. 9.5 A pair of welded sleeves on the submarine pipeline

welding to connect replacement sections to the original pipeline [9]. This method can be categorized based on the environment into wet welding, where the process takes place directly in water without isolation between the arc and the water, local dry welding, which creates a temporary dry area around the welding site, and dry welding, where water is entirely displaced to create a dry environment for the welding process [10, 11]. Wet welding is noted for its simplicity and direct application in water, albeit with potentially lower-quality outcomes due to the lack of arc isolation. Local dry welding and dry welding, on the other hand, aim to improve quality by excluding water from the welding area, with dry welding often occurring in a completely water-free environment, tailored to the surrounding pressure conditions. Dry hyperbaric welding is particularly suited for scenarios requiring high-quality welds to ensure pipeline integrity, despite its higher cost and the need for specialized equipment and divers certified in hyperbaric welding techniques. This method is essential for repairs where the pipeline cannot be surfaced and high-quality welding is crucial, necessitating complex systems, specialized equipment like hyperbaric welding chambers, underwater cutting tools, and support from heavy lifting vessels, alongside skilled divers with specific qualifications in dry hyperbaric welding. Figure 9.5 shows a pair of welded sleeves repairing the submarine pipeline.

9.2.2.5 Cladding Repair Method

The cladding repair method is an advanced technique used to maintain and enhance submarine pipeline integrity, particularly in high-corrosion environments such as offshore risers or areas severely affected by tidal and splash zone disturbances [12]. Various cladding materials can be used, with Stainless Steel 316 being the preferred choice due to its superior corrosion resistance. The key advantage of this method is its mechanical connection, which eliminates the need for welding and allows repairs to be conducted under operational conditions on platforms and risers.

Several critical steps are involved in implementing this method to ensure that the damaged area is effectively protected from corrosion while reinforcing the pipeline structure with cladding. Among these are sandblasting to prepare a clean surface for applying the cladding. The appropriate cladding material is selected and securely applied around the defective section of the pipeline. The cladding material may consist of stainless steel roll sheets or other commonly used materials, chosen for their resistance to harsh marine environments. The edges shall be sealed carefully with suitable sealing materials, and mechanical compression of rubber, or mechanical joints using rubber seals, shall be applied both at the bottom and top in combination to accomplish a watertight seal. After that, any water retained between the cladding and the pipe is removed, and the space is filled with heat- and corrosion-resistant resin to set the second protective coating. By eliminating welding, this method significantly reduces repair time for underwater pipelines. It is widely valued for extending the lifespan and enhancing the reliability of critical marine infrastructure.

9.2.2.6 Composite Repair

Composite repairs offer flexible and effective repair methods for a wide range of defects in submarine pipelines, benefiting from the application of composite materials due to their high strength-to-weight ratio, resistance to corrosion, and adaptability in underwater applications [13]. The materials used enable curing in wet conditions, thus providing enhanced structural and bonding integrity even under challenging marine conditions [5]. The selection of composite materials for submarine pipeline repairs is a critical decision that impacts the effectiveness, durability, and longevity of the repair solution. The choice of materials is guided by several factors, including the type of defect, the operating environment of the pipeline, and the mechanical and chemical properties required for the repair. Common composite materials used in submarine pipeline repairs include fiberglass for its cost-effectiveness and versatility, carbon fiber for its high strength and stiffness, and specialized resin systems that are designed to cure underwater, providing robust adhesion and corrosion resistance.

Various methodological approaches are applied depending on the type of submarine pipeline defect, including the use of these materials for full-length strengthening through wet lay-up systems or pre-impregnated wraps and coatings to reinforce areas weakened by corrosion, cracking, or leaks. The wet lay-up method involves wrapping the damaged section of the pipeline with composite fabric soaked in resin. Pre-impregnated wraps, already containing resin, are activated underwater to adhere to the pipeline's surface [13].

Determining the wrap thickness for submarine pipeline repairs involves assessing the pipeline's damage and mechanical properties, performing load and stress analysis, and calculating the required thickness based on industry standards, such as ASME PCC 2 2018 [14]. This process ensures the composite wrap can effectively redistribute stresses, withstand operational pressures, and meet industry standards for durability and safety. This standard enables the repair of metallic pipelines using composite overwraps for different repair scenarios, including hoop stress, axial stress, and leak mitigation. It incorporates the ASME strategy of accommodating the yielding (plastic deformation) of metallic pipes in the repair strategy. Concurrently,

ISO 24817 outlines a procedure similar to that of ASME PCC-2-2018, yet it distinctively excludes pipe yielding, making it applicable for the repair of non-metallic and brittle pipes [15].

For larger structural repairs, sleeve systems such as split sleeve repair and Cured-in-Place Pipe (CIPP) techniques are used [16]. Split sleeve repair involves encasing the pipeline section with a prefabricated composite sleeve and sealing it with adhesive resins to repair cracks or leaks. The CIPP method rehabilitates pipe interiors by inserting a resin-impregnated liner into the existing pipeline and curing it in place to form a new, structurally sound inner shell.

Localized damages and heavy corrosion are addressed through patch repairs and full encapsulation, respectively. Composite patches, pre-impregnated with resin, offer quick and localized repair solutions for small areas of damage. Full encapsulation involves wrapping the entire affected section with multiple layers of composite material to provide extensive structural reinforcement with a new load-bearing outer shell.

The choice of repair method is based on the detailed nature of the pipeline defect, operational conditions, and repair objectives, ensuring that the most economical, efficient, and reliable repair solution is applied to maintain the integrity and functionality of submarine pipelines.

9.2.2.7 Robotics in Submarine Pipeline Repair

Robotic technologies like COVERCAT's Pipe Snake and the Nautilus robot are revolutionizing submarine pipeline repair, presenting cutting-edge solutions for the upkeep and restoration of pipeline integrity. The Pipe Snake, focusing on internal pipeline repairs, adeptly navigates within pipes to directly apply protective coatings to the internal surfaces. This sophisticated robot is capable of detecting and filling holes, as well as grinding and preparing surfaces for coating, effectively combating internal corrosion and roughness to improve flow rates.

In contrast, the Nautilus robot represents a significant advancement in external pipeline repair. A collaboration between ROVULA and Kongsberg Ferrotech has resulted in Nautilus, a pioneering force in subsea pipeline inspection and repair robotics. It specializes in conducting NDT to identify defects and carry out repairs on damaged pipeline sections. The Nautilus system is equipped to handle a wide range of repair tasks, from defect assessment and surface preparation to the application of structural reinforcements like carbon fiber, all with robotic precision [17]. Impressively, Nautilus can significantly reduce repair durations from weeks to under 48 h, cutting down preparation costs by more than 50%.

These robotic systems exemplify the vast potential of automation in improving the efficiency, safety, and environmental sustainability of submarine pipeline repair operations. By adopting robotic technologies, the oil and gas industry can secure more dependable and cost-efficient maintenance procedures, ensuring the extended functionality of essential underwater infrastructure while reducing the risk to human workers in potentially dangerous conditions.

9.3 Emerging Technologies and Future Trends

The landscape of submarine pipeline repair is undergoing a significant transformation, driven by technological advancements in robotics, smart materials, digital twin technology, and a stronger emphasis on environmental and regulatory factors. Robotics and Autonomous Underwater Vehicles (AUVs), such as the Nautilus robot and COVERCAT's Pipe Snake, are revolutionizing inspection, maintenance, and repair operations. These technologies facilitate precise assessments and repairs in areas that are challenging to access, eliminating the need for direct human involvement and markedly decreasing both the time and cost associated with operations.

At the same time, the development of smart materials offers a range of solutions for repairing submarine pipeline defects, from self-healing capabilities that automatically address cracks and microfractures, to shape memory alloys and corrosion-resistant coatings that provide structural support and protection against environmental damage. The integration of these materials into repair strategies not only enhances the durability and reliability of submarine pipelines but also contributes to more sustainable and cost-effective operations in the offshore industry.

Integrating digital twin technology with machine learning revolutionizes submarine pipeline maintenance, elevating safety and operational efficiency to unprecedented levels [18]. The major process of using digital twin technology for maintenance is shown in Fig. 9.6. Through real-time data from sensors, this synergy provides a comprehensive, actionable view of the pipeline's condition, enabling immediate responses to deviations. Machine learning goes further, forecasting potential issues and optimizing maintenance schedules to preemptively address problems, significantly reducing emergency intervention costs. This approach also facilitates the simulation of repair scenarios, enhancing decision-making for complex deep sea operations. As this technology evolves, it continually improves maintenance strategies and informs future pipeline designs, promising a future where submarine pipeline operations are not only more sustainable and resilient but also intelligently proactive in ensuring efficiency and safety.

Furthermore, the industry's shift towards environmentally conscious and regulation-compliant technologies is becoming increasingly pronounced, underscoring a commitment to sustainable development. This transition is characterized by adopting biodegradable materials that significantly reduce environmental footprints. For instance, using biodegradable hydraulic fluids in underwater equipment and ROVs for repair operations exemplifies this trend. These fluids are engineered to decompose rapidly in marine environments, mitigating pollution risks. Similarly, eco-friendly coatings, such as ceramic coatings, and epoxy coatings with organic inhibitors, for pipelines represent another stride towards sustainability, offering protection against corrosion while being non-toxic to marine life. These innovations are guiding the industry towards more effective, reliable, and eco-friendly repair techniques, with the prospect of fully autonomous repair systems representing a significant breakthrough. As these technologies continue to evolve and

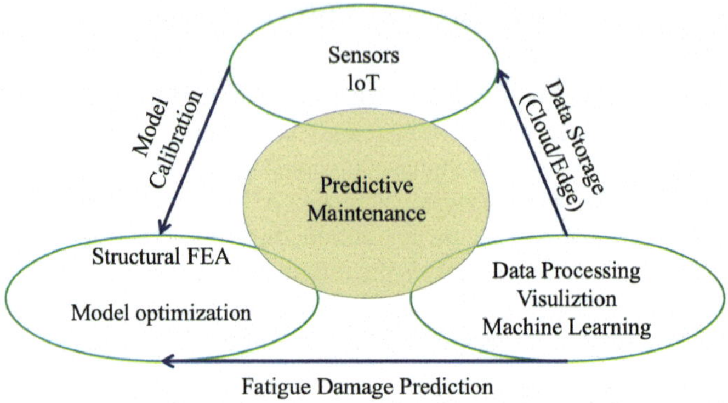

Fig. 9.6 The major process of using digital twin technology for maintenance

become more integrated, they promise to fundamentally change the approach to submarine pipeline maintenance, ensuring operations that are both safer and more environmentally sustainable.

References

1. Popineau D, Wiet P, Boulet d'AURIA S (2012) Subsea pipeline repair by composite system. OnePetro
2. Andersen T, Misund A (1983) Pipeline reliability: an investigation of pipeline failure characteristics and analysis of pipeline failure rates for submarine and cross-country pipelines. J Petrol Technol 35(04):709–717. https://doi.org/10.2118/10467-PA
3. Zhao T-F, Zhang Q-Y, Chu G (2021) Truss-type maintenance devices and corresponding pipeline lifting control strategies. Pet Sci 18(3):939–950. https://doi.org/10.1007/s12182-021-00545-2
4. Zhao Z, Shi X, Ma P (2023) Analysis on contact pressure of the seal structure of repair clamp for subsea pipeline. OnePetro
5. Noor N, Azraai SNA, Lim KS, Yahaya N (2015) An overview of corroded pipe repair techniques using composite materials. Int J Chem Mol Nucl Mater Metall Eng 10:19–25
6. Cao Y, Zhang L, Sun Y, Shi Y, He Y, Zhen Y (2020) Strength analysis of the interference joint of the oval submarine pipeline to the mechanical connector. Appl Ocean Res 101:102229. https://doi.org/10.1016/j.apor.2020.102229
7. Haider S, Sen M, Lawrence D, Rodayan A (2021) Integrity management of flange connections using reliability model. Am Soc Mech Eng Digit Collect
8. He Y (2018) Study on mechanics problems in underwater repair of subsea pipeline. Master's Thesis, China University of Petroleum (East China)
9. Yin Y, Yang X, Cui L, Cao J, Xu W (2015) Investigation on welding parameters and bonding characteristics of underwater wet friction taper plug welding for pipeline steel. Int J Adv Manuf Technol 81(5):851–861. https://doi.org/10.1007/s00170-015-7281-1
10. Rogalski G, Fydrych D, Łabanowski J (2017) Underwater wet repair welding of API 5L X65M pipeline steel. Pol Marit Res 24(s1):188–194
11. Wang K, Chen J, Huang J, Lin P, Guo J, Zhu H, Fang Y, Chen X, Cao G, Jia W, Gao X, Ge H (2024) Investigation of dry environment construction equipment's medium displacement

system for submarine pipelines. Ocean Eng 300:117365. https://doi.org/10.1016/j.oceaneng.2024.117365
12. Subhan A, Zahoor A, Mourad A-HamidI (2022) Advances in manufacturing techniques of cladding steel pipes using corrosion-resistant alloy material for offshore oil and gas pipelines. In: 2022 Advances in science and engineering technology international conferences (ASET), pp 1–6
13. Shamsuddoha M, Islam MM, Aravinthan T, Manalo A, Lau K (2013) Effectiveness of using fibre-reinforced polymer composites for underwater steel pipeline repairs. Compos Struct 100:40–54. https://doi.org/10.1016/j.compstruct.2012.12.019
14. Ortiz X, Jungwirth D, Behnamian Y, Sharahi HJ (2021) Composite sleeve repair in the North American regulatory environment. Am Soc Mech Eng Digit Collect
15. Saeed N, Ronagh H, Virk A (2014) Composite repair of pipelines, considering the effect of live pressure-analytical and numerical models with respect to ISO/TS 24817 and ASME PCC-2. Compos Part B: Eng 58:605–610. https://doi.org/10.1016/j.compositesb.2013.10.035
16. Hicks J, Kaushal V, Jamali K (2022) A comparative review of trenchless cured-in-place pipe (CIPP) with spray applied pipe lining (SAPL) renewal methods for pipelines. Front Water 4. https://doi.org/10.3389/frwa.2022.904821
17. Carlsen C, Nagadhana T (2024) The nautilus robotic solution: a single robot that integrates inspection + maintenance + repair capabilities for subsea pipelines and tubulars. OnePetro
18. Chen B-Q, Videiro PM, Guedes SC (2022) Opportunities and challenges to develop digital twins for subsea pipelines. J Mar Sci Eng 10(6):739. https://doi.org/10.3390/jmse10060739

The manufacturer's authorised representative in the EU is Springer Nature Customer Service Centre GmbH, Europaplatz 3, 69115 Heidelberg, Germany. If you have any concerns regarding our products, please contact ProductSafety@springernature.com

Printed and bound by CPI Group (UK) Ltd, Croydon, CR0 4YY
05/10/2026
02228127-0001